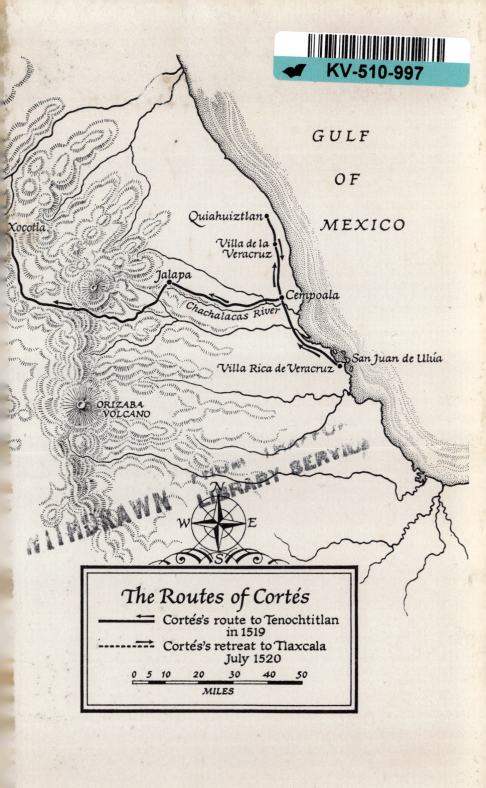

GULF

OF

MEXICO

Xocotla

Quiahuiztlan

Villa de la
Veracruz

Jalapa

Cempoala

Chachalacas River

Villa Rica de Veracruz

San Juan de Ulúa

ORIZABA
VOLCANO

N
W   E
S

The Routes of Cortés

→ Cortés's route to Tenochtitlan
in 1519

---→ Cortés's retreat to Tlaxcala
July 1520

0 5 10   20   30   40   50
MILES

Cortés

Anonymous portrait of Cortés, from the Hospital de Jesús,
Avenida 20 Noviembre, Mexico City. Photo by Juan Guzmán

Cortés

William Weber Johnson

THE LIBRARY OF WORLD BIOGRAPHY
GENERAL EDITOR: J. H. PLUMB

HUTCHINSON OF LONDON

Hutchinson & Co. (Publishers) Ltd
3 Fitzroy Square, London W1

London Melbourne Sydney Auckland
Wellington Johannesburg and agencies
throughout the world

First published in Great Britain 1977
© William Weber Johnson 1975
Maps © Little, Brown & Company 1975

Acknowledgements

The author is grateful to the following publishers for permission
to quote selections from previously copyrighted materials:

Grossman Publishers, for *Hermán Cortés: Letters from Mexico*,
translated and edited by A. R. Pagden, copyright © 1971 by
A. R. Pagden; and for *The Conquistadors*, edited and translated by
Patricia de Fuentes, copyright © 1963 by The Orion Press, Inc.

Doubleday & Company, Inc., for *The Bernal Díaz Chronicles,
The True Story of the Conquest of Mexico*, translated and edited by
Albert Idell. Copyright © 1956 by Albert Idell.

The Cortés Society, for *The Rise of Fernando Cortés* by Henry R.
Wagner. Copyright 1944 by the Cortés Society.

The Regents of the University of California, for *Cortés: The
Life of the Conqueror by His Secretary* by Francisco López de Gómara,
edited and translated by Lesley B. Simpson and originally
published by the University of California Press. Copyright
© 1964 by The Regents of the University of California.

Printed in Great Britain by litho at The Anchor Press Ltd
and bound by Wm Brendon & Son Ltd
both of Tiptree, Essex

ISBN 0 09 128480 5

For my father
Finis Ewing Johnson, 1878–1941,
who loved both history and adventure

Consider his ingratitude . . . his double and deceitful dealings
. . . his treachery . . . the useless massacre . . . his insatiable
desire for gold and for pleasures. . . . Accuse him of everything else which history records as proven. But then allow him
the plea that he was a sagacious politician and a valiant and
able captain; that he concluded successfully one of the most
astounding feats of modern times. . . . From the midst of such
opposite elements we see the gloomy and noble figure of Don
Hernando rise head and shoulders above the common stature
of humanity.

— MANUEL OROZCO Y BERRA

Introduction

WHEN WE LOOK back at the past nothing, perhaps, fascinates us so much as the fate of individual men and women. The greatest of these seem to give a new direction to history, to mold the social forces of their time and create a new image, or open up vistas that humbler men and women never imagined. An investigation of the interplay of human temperament with social and cultural forces is one of the most complex yet beguiling studies a historian can make; men molded by time, and time molded by men. It would seem that to achieve greatness both the temperament and the moment must fit like a key into a complex lock. Or rather a master key, for the very greatest of men and women resonate in ages distant to their own. Later generations may make new images of them — one has only to think what succeeding generations of Frenchmen have made of Napoleon, or Americans of Benjamin Franklin — but this only happens because some men change the course of history and stain it with their own ambitions, desires, creations or hopes of a magnitude that embraces future generations like a miasma. This is particularly true of the great figures of religion, of politics, of war. The great creative spirits, however, are used by subsequent generations in a reverse manner — men and women go to them to seek hope or solace, or to confirm despair, reinterpreting the works of imagination or wisdom to ease them in their own desperate necessities, to beguile them with a sense of beauty or merely to draw from them strength and understanding. So this series of biographies tries in lucid, vivid, and dramatic narratives to explain the greatness of men and women, not only how they managed to secure their niche in the great pantheon of Time, but also why they have continued to fascinate subsequent generations. It may seem, therefore, that it is paradoxical for this series to contain living men and women, as well as the dead, but it is not so. We can recognize, in our own time, particularly in those whose careers are getting close to their final hours, men and women of indisputable greatness, whose position in history is secure, and about whom the legends and myths are beginning to sprout — for all great men and women become legends, all become in history larger than their own lives.

Only Spain in the sixteenth century could have bred Cortés — poor yet arrogant, chivalrous yet cruel, pious yet sinful, generous yet full of greed; above all, a man of decisive action, but capable of patience and guile, a man who belonged to a world that had passed yet who launched half a continent and its people into the mainstream of history. Had Cortés been a failure, had he never stirred beyond his native Extremadura, so long as he had left behind his own literary evidence, he would have lived in history, for his pen was as sharp and as vivid as his sword. But unbelievable success came to Cortés: with a handful of men and a few horses he toppled the greatest, the most warlike of all American empires — the Aztecs of Mexico.

The triumph of Spain in the early sixteenth century in the New World is one of the great events of Western history; one that has changed the world. For centuries Spain had been involved in aggressive crusading war, steadily forcing back the

Moors who had flooded into Spain in the eighth century. The fighting was bitter, prolonged, interspersed with long, uneasy truces. As Spain began to succeed, so did its religious fervor grow, breeding a sense among knights who undertook the struggle that they were the chosen instruments of God, the spearhead of faith. And it is impossible to understand the triumph of Cortés and his captains if one ignores the depth and sincerity of their religious beliefs, no matter how brutal and sensual they might be in action. They could torture, kill and rape, but they knew how to pray to the Virgin and the saints.

The great heartland of Spain, where Cortés was born, was barren and poor, a plateau of limitless, windswept horizons that made neither nobleman nor peasant rich. For the *hidalgos,* the gentlemen of birth and pride, there was in this land little hope of much but dignity and poverty; both the south and east of Spain were richer, more thriving, more attuned to the world's changes in trade and exploration. The discovery of Española, the news of riches and gold, worked like yeast on the dreams of men such as Cortés and Pizarro, or even on humble soldiers such as Bernal Díaz. It is not surprising, therefore, that many of the conquistadors were drawn from the arid heart of Spain, to Seville, to Cádiz, and then to the Indies. There lay the charm of riches; there lay the hope of fame by the sword. Glory and gold, they longed for both. They longed also to serve God and their King. The whole of the great Spanish empire was a curious mixture of crusading zeal and heartless greed. Cortés was full of lust — for women, for riches, for power. Nevertheless he was deeply pious, certain of Heaven, careless about death, hence capable of a daring that still amazes.

Even Cortés, however, must have been astonished by his own success as his followers were astounded by the civilization which they discovered — as strange, as unreal, as dreamlike as the discovery of a new planet would be to us. Neither he nor his followers had seen so vast a city as Mexico, or one so teeming with people, with food, with riches of every kind. Nor one so barbarous; so addicted to a religion of blood. The art of the Aztecs is a terrible art, and reflects one of the most savage religions known to mankind, in which thousands of men and women, boys and girls were ritually slaughtered. Even conquistadors as tough as Cortés, used to quick and violent death as they were, became appalled by the slaughter and by the cannibalism. Yet the terrible nature of that religion was to prove their greatest ally, greater, far, than the strange portents, the fearful phantoms which paralyzed the will of Montezuma, the godlike King of the Aztecs. The endless sacrifices bred deep hatred in the tribes subject to the Aztecs, and so made them eager allies of Cortés. Without them his tiny band could never have toppled the Aztec empire with such relative ease.

Although often told, the tale of Cortés is of perennial fascination, partly because Cortés himself is so complex and tragic a character, and partly because few men of action have changed so profoundly the destiny of a continent. Cortés's conquest made him enormously rich and powerful — the Marquis of the Valley, with thousands of slaves — indeed, an estate as huge as a kingdom — and yet nemesis overtook him. Returning to Spain, he was kept there by his hopes, denied the honors he felt were rightly his. He lived in darkening despair, on into a bleak old age of cantankerous litigation. Tragic and sad it might be for Cortés; the fate of the Mexicans was more terrible. The diseases — particularly smallpox — brought in by the Spaniards decimated the population. The ruthless savagery of their new conquerors killed far more than the sacrifices demanded by the Aztecs, and one of the most populous and fertile valleys of the world lost both its people and its vitality. The Christian city of Mexico, built on the ruins of the old, was but a shadow of the former greatness. Mercilessly exploited, totally subjected, Mexico became Spanish and Christian; irradicably so, no matter how blended with cultural strains from the Aztec past.

Cortés, of course, is no longer a Mexican hero — the stress, rightly, is on a revival of Indian culture, of Indian self-confidence in which there is little room except for denigration of the conquistadors. Yet the result of their conquest still reverberates — Spanish is the language, and at the great shrine of Guadalupe one may see every hour of every day the rapt Indian faces, lost in prayer to the Madonna.

— J. H. PLUMB

Contents

Cortés

ONE

Extremadura

By the time he was forty-five Hernán Cortés, the conqueror of Mexico and creator of New Spain, had become a *don,* a grandee of Spain, and, as Marquis of the Valley of Oaxaca, the ruler and beneficiary of a vast and rich part of the New World. And he had adopted a coat of arms suitable to his recently attained high station.

Its motto: "The judgment of the Lord overtook them, and His might strengthened my arm."

It really was not that simple.

The exotic native dominions that Cortés overwhelmed, sacked and ground into the dust of the New World were as much doomed by their own frightful, demanding deities as they were by the one Cortés claimed as his divine partner in the bloody enterprise.

And his strong arm, with or without heavenly assistance, was not his principal weapon. His guile, his wit, his quick perception of others' weaknesses and strengths, his intuitive sense of how best to use them were all more potent in his daring adventure. So was his determination, his ruthlessness in achieving his ends. So were his addiction to gambling and, more significantly, his remarkable luck.

And there was also the portentous fact of his time and origin. Europe's constant and confusing wars of the late Middle Ages had produced a military phenomenon: the military leader who functioned not only as strategist, tactician and battlefield captain, but also as chief justice, executioner, diplomat and governor of occupied territory. In some cases his sovereign might choose to lead his men in person. More often the sovereign's judgments, strategic decisions, were conveyed to a distant field of action by couriers who might die or disappear or even lose or forget or misconstrue the message before it could ever be delivered. Because of the difficulty of communication the captain in the field was on his own and became a de facto sovereign himself. He held in his own hands, whether capable or otherwise, destinies of peoples and nations — and also the ingredients of his own glory. Or ignominy.

Cortés was born in the last years of the Middle Ages and he came of age just as Spain's *siglo de oro,* its golden century, was beginning. He was an exemplary Spaniard whose life and career would span the high tide of Spain's imperial grandeur, the greatest empire since Rome.

Being exemplary was not a simple matter. Spaniards of the fifteenth and sixteenth centuries were an astonishing mix of bloods and cultures, the end product of nearly two thousand years of tribal wars, invasions, rebellions, internecine rivalries, crusades, conquests and reconquests. To the character, customs and attitudes of the ancient indigenous Iberians had been added those of the Celts, Phoenicians, Greeks, Carthaginians, Romans. The centuries of Roman domination had tested and developed the Iberian peninsula's human resources. From Spain came a long line of fierce and fearless legionaries, distinguished intellectuals such as the two Senecas and Martial, and a number of "Roman" emperors — Constantine, Hadrian, Marcus Aurelius, Trajan, Theodosius.

As Rome decayed the Visigoths took over. Their primitive Christianity strengthened Catholicism, which was already

making rapid strides in Spain. And they introduced monar-
chical government, which was to be a key element in Spanish
life thereafter. From the Visigoths probably came the fair skin
and hair of many of the conquistadors of the New World,
complexions that were to play a curious role in history.

And finally there were the Moslems, who controlled the
destinies of the Iberian peninsula for almost eight centuries.
The "Moors," as the Spaniards inaccurately described the
Moslem invaders, introduced hitherto unknown arts, sciences
and skills. They left indelible marks on education, architec-
ture, poetry and language and contributed substantially to the
traits of Spanish character.

Spaniards of the sixteenth century had many of the mark-
ings of a master race: towering pride; hypersensitivity in
matters of honor; vengefulness in either real or imagined
wrongs; individualism that often differed little from anarchy;
religious fervor that could excuse almost any excess committed
under the proper auspices; cruelty in battle and mercilessness
with victims; a disdain for menial labor and, above all, a
consuming lust for gold or other spoils.

The long struggle against the Moslems had brought about
the greatest unity Spain had ever known: the joint rule of the
Catholic monarchs, Ferdinand and Isabella. Under them the
Moorish domination of Spain ended with the fall of Granada
in 1492. In the same year the visionary navigator from Genoa,
Christopher Columbus, with some help from Queen Isabella,
sailed west across the Atlantic to what he thought were the
Indies. His discovery was to change the history of the world in
a far more significant way than had the fall of Granada.

Both events occurred in the eighth year of the life of Hernán Cortés, a man who was to make his own radical changes
in the nature and boundaries of the known world. He was
born in Medellín, on the banks of the Guadiana River in the
province of Extremadura in 1485. In addition to being a child
of the new Spain that was coming into being he was also an
extremeño, a native of the province which the Romans had

described as not only remote and hard (hence the name) but also horrid and bellicose. The windswept steppes of Extremadura, broken only by rolling mountains and groves of cork oaks and olive trees, were a region of poverty and bleakness, of blinding heat and bitter cold, thin soil, meager resources and comforts. It was the kind of place that inspires men to go on risky ventures in far parts of the world in the hope of finding wealth, power, glory or, all else failing, at least an improvement on what they had left behind. Significantly, no other province in Spain produced so many conquistadors in the early stages of the exploration and exploitation of the New World: Cortés of New Spain, or Mexico; Vasco Nuñez de Balboa of Darien, or Panama; the Pizarros of Peru; Pedro de Valdivia of Chile; Hernando de Soto — many more.

The *extremeños* were, in large part, descendants of a people the Romans had found to be unruly, unimpressed with the grandeur of Rome and disinclined to subjugation. On various occasions they astonishingly defeated large, well-equipped Roman armies in the field. These rude frontiersmen did not have enough sense to know when they were beaten. They simply kept on fighting, no matter how great their losses. A shepherd, Viriatus, waged guerrilla warfare on the Romans in the second century B.C. and became one of Spain's great military heroes. For many years Viriatus fought off the Roman legions, darting in and out of the mountains of Extremadura inflicting disasters on an enemy that was superior in numbers, training, experience and weapons. Viriatus was so invincible that in the end the Romans could deal with him only through treachery. Some of Viriatus's own men were persuaded by the taste of Roman wine and the heft of Roman gold to kill the peasant hero by stealth. There was betrayal as well as courage in the heritage of Extremadura.

Another example of the spirit of Extremadura came almost six centuries later. By that time Christianity was spreading rapidly in the remote province. Children and young people were particularly susceptible to the teachings of Christ. On one occasion a Roman official in Mérida — which was said to be the ninth city of the Roman empire — was publicly denouncing Jesus and all his followers. One of his listeners, a young girl named Eulalia, answered by spitting in his face. In punishment she was burned at the stake, somewhere near the site of the future basilica of Santa Eulalia in Mérida.

How deeply rooted the Cortés family was in Extremadura is uncertain. The name may have come from the Lombardic Cortesio. The family may have originated in Aragon. Or possibly Salamanca. But, even with this vagueness of antecedents, the Cortés family, of which Hernán was the only child, was one of some substance. Cortés was considered an hidalgo (from *hijo de algo,* son of something, as opposed to a nobody). His mother, remembered as a very pious but also a hard and stingy woman, was Doña Catalina Pizarro Altamirano, the daughter of the majordomo of the Countess of Medellín. She was always addressed with the honorific title of *doña.* The comparable masculine form, *don,* was never accorded to the father, Martín Cortés de Monroy. But, said Francisco López de Gómara, one-time secretary to Cortés and later his biographer, "the four lines of Cortés, Monroy, Pizarro and Altamirano are very ancient, noble and honorable. They had little wealth but much honor, a thing that rarely occurs except among those of virtuous life." Property belonging to Doña Catalina brought in a small income from honey, wheat and wine and with this the Cortés family had a life of modest gentility.

The father, said to be both devout and charitable, was an old soldier, as were so many Spaniards in those troubled times. He had served in a company of cavalry of the Order of Alcántara, a military-religious organization which owned about half of the province of Extremadura. Queen Isabella, in the interest of unifying her realm, asserted her supremacy over the various autonomous regional orders that had, until then, been laws unto themselves. Some of the orders, like that of Alcántara, resisted — uselessly, as it turned out. The struggle had given Martín Cortés experience under arms. This probably was a factor in what must have been a chivalric ambience in the Cortés household. If Cortés had an ideal as a youth it almost certainly would have been the Great Captain, Gonzalo Fernández de Córdoba, Spain's great military luminary in the struggles with the Moors and the French.

As a child Cortés was sickly and rachitic. His nurse, Mariá de Esteban, fearing for his life, was said to have written the names of the twelve apostles on slips of paper and put them in a jar. She then drew one to see which should be the child's patron. It was St. Peter and she said special prayers to him for the child's health (and in later life Cortés always observed St. Peter's feast day when he could). This early frailty may have precluded any planning for a military career.

One version of Cortés's youth is that he served as an acolyte in a church in Medellín where he picked up a familiarity with the Psalms (which he often quoted later in life), theology and Latin, and that he later worked as an *escribano público,* or notary public, in Salamanca, Valladolid and Seville before embarking for the Indies.

A more detailed and colorful version is generally accepted. At fourteen the boy was sent off to the great university at Salamanca, one of the four most important centers of learning of the Christian world. There he studied Latin, grammar and law. He must have done well. Later in his career his comrades admired his skill at conversing in Latin, his ability to compose forceful *relaciones,* or dispatches, legal or at least legal-sounding documents and even poetry. Whatever he picked up at the university he picked up quickly, for at sixteen he was back in his parents' home in Medellín either because of illness (he periodically suffered from quartan fever) or just plain boredom.

López de Gómara says that at this time Cortés was "restless, haughty, mischievous and given to quarreling." He must also have been a trial to his parents. It was decided that he should go to seek his fortune or, more bluntly, get out. There appeared to be two ways to go. One would be to join the Great Captain in Naples for his campaign against the French. The other would be to go to the Indies, which now, just at the turn of the century, were drawing more and more adventurous young men from Spain. Returning conquistadors, with golden chains around their necks, pockets and pouches filled with gold and jewels, trains of naked or gaudily dressed Indians, crates and cages of exotic birds and beasts, strange banners and clouds of smoke from a fragrant weed called tobacco, were creating a heady atmosphere. Each returning ship brought additional knowledge of the world Columbus had found, but it was clear there was much more to be discovered, much more to be exploited.

So, for Cortés and his family, the voyage to the west was an easy choice. In addition, Nicolás de Ovando, under whom Martín Cortés had served in the Order of Alcántara, was going out to the Indies as governor general. It was arranged for young Cortés to go out with him.

But, as it was so often to do later in his career, Cortés's fascination with and for women interfered. While Ovando's fleet was being readied for departure Cortés literally fell into trouble. Returning one night from a love tryst, he was stealthily making his way along the top of a wall. The masonry crumbled under his feet and he crashed into a garden. An irate husband stormed out of the house and would have run the intruder through with his sword had it not been for the intervention of the swordsman's mother-in-law. But Cortés, while he escaped the sword, was injured in the fall. He took to his bed and had another bout with quartan fever. Meanwhile Ovando's fleet sailed without him.

Recovered, young Cortés decided to join the Great Captain in Italy after all. He set out for Valencia, a port of embarkation. He did not get to Italy. Instead, apparently penniless and aimless, he simply wandered around Spain. The supposedly easy wealth of the Indies became more and more alluring. Finally he returned to Medellín and his parents raised the money necessary for him to get to the Indies. Money was short, but so too must have been their patience by this time.

With a little baggage and a supply of preserves and jams from home Cortés boarded a merchant ship in Sanlúcar de Barrameda early in 1504. The ship was one in a convoy of five merchant vessels carrying supplies to the Spanish island outpost of Española (the island now occupied by Haiti and the Dominican Republic).

The first part of the trip was a sort of shakedown cruise to La Gomera in the Canary Islands. In this way pilots and crews had some sea experience before beginning the more difficult part of the journey across the Ocean Sea. It had only been a dozen years since Columbus had sailed out into this terrifying vastness. Although it had been demonstrated that the world was round, there were still some who feared falling off the edge of the earth.

At La Gomera the ships took on fresh food and water. Crews and passengers said prayers for a safe passage. The master of the ship on which Cortés was sailing, Alonso Quintero, began to exhibit the sort of unbridled individualism for which the Spaniards were to become notorious in the New World. While the other four ships were still completing arrangements for departure, Quintero hoisted sail and left, hoping to gain an advantage of a day or so in reaching Española and putting his cargo on sale.

But before Quintero's ship had passed the outermost of the Canary Islands it was struck by a gale that carried away the mainmast, covering the deck with broken rigging and shredded sails. Quintero managed to work his crippled ship back to La Gomera. There he pleaded with the other ship captains to wait while the needed repairs were made to his vessel. Surprisingly, they agreed.

Finally all ships were ready and an orderly departure was made. But, once at sea, Quintero tried once more to gain an advantage over his companions. When a brisk wind came up one night Quintero ordered all sail piled on and once more sailed away from the convoy.

But either Quintero or his pilot, Francisco Niño, one of the seafaring family that had served Columbus, lost the way. They blamed each other. Whoever was at fault, they were lost. After a time food and water began to run short. Cortés's personal stock of jams and preserves was long since exhausted. So were the ship's stores of moldy bread and dry meat. Rainwater from the sails was caught and hoarded. And there was nervous talk of dreadful alternatives: starvation at sea or shipwreck on the islands of the Indies, where the natives were said to eat human flesh.

But on Good Friday — a day that was always to have special significance for Cortés — a white dove circled the mast, lighted on a yardarm and then took off. The ship followed in the same direction. Soon land was sighted. The pilot, Niño, with more certainty than he had displayed for many days, identified it as the shore of Española.

When Quintero's ship finally worked its way into the harbor at Santo Domingo on Easter Sunday the master's relief was tempered by chagrin. The other four ships, which he had tried to leave at sea, were already there and doing business.

There was a lesson in this for any young man setting out to conquer the world.

TWO

The Indies

ESPAÑOLA, THE FIRST European settlement in the Americas, was the administrative center for Spain's new overseas empire. The island was lush and fertile. Seeds and cuttings brought from Spain flourished, and livestock multiplied prodigiously.

Still, the colony was frequently on the verge of starvation, largely because of the Spanish immigrants' disdain for manual labor. The Great Admiral, Columbus, although he continued to praise its virtues extravagantly, was not only disappointed by the colony he had founded; he was also virtually destroyed by it.

He had touched the island on his first voyage and had brought a thousand or so colonists for it on his second. It did not fare well. The colonists were interested in gold, little else, and there was not much of it in Española. The Spanish settlers, insubordinate by nature, disliked and distrusted the Italian-born Columbus. The latter, in desperation, enforced rationing of scarce food supplies and decreed that in order to eat everyone must work — a disagreeable sort of proposal.

On his third voyage, the one on which he touched the South American mainland for the first time, Columbus found

Española in a deplorable state. He had left his brother Bartolomé in charge. The town of Santo Domingo had been built, but little else had been done in developing the colony. Instead, a state of near-anarchy prevailed. Columbus restored order as well as he could by making large grants of land to the rebellious colonists and by conceding the right to exact forced labor from the natives. This was the beginning of the odious system of *repartimientos*, literally distributions of human beings. The indigenes were treated as nothing more than a commodity — and a perishable one, since overwork and European diseases soon reduced them in number. It was a system that would blacken Spain's reputation in the century of her greatest triumphs as a nation.

Stories of Columbus's supposed misconduct and his alleged mistreatment of both Spaniards and natives were brought back to Spain by disappointed homeward-bound colonists, many of whom took Indian slaves with them. Queen Isabella was shocked. Although she and King Ferdinand continued to believe in Columbus, they sent out a commissioner to look into the affairs in Española, the extent of slavery and the failure to Christianize the natives. Their appointee was Don Francisco de Bobadilla, master of the knightly order of Calatrava. On the queen's orders several hundred of the recently arrived slaves were liberated and repatriated on the same ships that carried Bobadilla and his party.

Bobadilla had supreme powers to restore peace and order in the colony. His first act was to seize the Great Admiral, put him in chains and send him back to Spain. Upon arrival in Spain, Columbus was immediately freed on the monarchs' orders and reinstalled in his position of honor. He was also given authority for his fourth voyage of discovery in order to find a westward passage to the Orient. But at the same time he was specifically ordered to stay away from Española.

Bobadilla did little more than aggravate the unrest that had existed under Columbus and his brother Bartolomé. He was replaced in 1502 by Nicolás de Ovando, the old friend of the

Cortés family. Ovando was both a soldier and a religious friar, a man of stern righteousness. He imprisoned Bobadilla and those of the colonists who were loyal to him and prepared to send them back to Spain. He alleviated many of the disorders, although he displayed a certain callousness toward the abused natives.

Cortés, when he arrived in Española in 1504, carried a letter of introduction to Ovando, but Ovando was away from Santo Domingo at the time. Cortés met, instead, the governor's secretary, a man named Medina. The nineteen-year-old adventurer was informed of his rights and perquisites as a conquistador. He was entitled to a grant of land for farming, a building lot for a house and a parcel of Indians to do the work (Queen Isabella's horror of slavery had not eradicated it). Five years of residence and industry would confirm his rights. After that he might go and do as he liked. Cortés is said to have replied: "Neither in this island nor in any other part of the New World do I wish or mean to stay so long."

Nevertheless, he accepted the offered land, put his Indians to work looking for gold — futilely, as it turned out — and became notary public in the village of Azua, some fifty miles west of Santo Domingo. It was a task for which he had some preparation, but it paid little. He had no money, no resources, and he and two other young Spaniards took turns on the use of a single cloak for appearances in public (no proper Spaniard would be seen in public without his cloak, even in this tropical climate).

And, despite his brash statement to Medina, Cortés contented himself for seven lean years in Española — and this at a time when the island was a staging area for tantalizing expeditions fanning out in all directions. Some were slave hunters; the native population (that is, the work force) of Española was dwindling rapidly. And many were explorers looking for other promising corners of the New World.

In 1509 — by this time he had fulfilled the five-year residence requirement — Cortés had been tempted to join the expedition of Diego de Nicuesa to establish a colony in Veragua in what is now Panama. The expedition ended in disaster. But it was not foresight that had kept Cortés from joining it. Instead, it was his old trouble: his love life. At the time of departure he was suffering from a venereal infection, probably syphilis, which the Spaniards complained they frequently contracted from Indian women. They treated it with frequent infusions made from lignum vitae, called *guayacán* by the natives and *palo santo*, or holy wood, by the Spaniards. In addition, Cortés had suffered at least one knife wound in a fight over a woman (the chronicler Bernal Diáz del Castillo said the scar of this encounter was on Cortés's chin, usually covered with a beard).

During his recovery from one of these mishaps Cortés is said to have had a prophetic dream. He found himself wearing rich garments and being served by strange people who addressed him in terms of great honor and deference. Although he professed to take little stock in dreams he told friends about it. And he added the prediction that in time he would either "dine with trumpets or die on the gallows." It seemed to signal the end of Cortés's poverty and obscurity.

By this time Spain's little seagirt empire was expanding. Ovando had succeeded in stabilizing Española — at considerable cost to the native population — and had sent Juan Ponce de León to subjugate and exploit Boriquén, the island to the east, now Puerto Rico. Ovando was replaced by Diego Colón, the son of the Great Admiral, who had died a bitter and disappointed man in 1506. Diego Colón had inherited his father's title as viceroy of the Indies, and he was free to push back the boundaries. He sent out an expedition to explore and colonize Jamaica to the west. By 1511 he was ready to take over Cuba, to the northwest. His father had mistakenly insisted that Cuba was a peninsula extending out from the mainland of Asia. By this time it was known that Cuba was an island and apparently a large and fertile one. But little more was known.

As leader of the expedition Colón chose Diego Velázquez, a veteran of the Spanish wars and a pioneer settler in Española — he had come out in Columbus's second voyage. He was a prominent and prosperous citizen of Española and was the founder of the towns of Yaquimo, Maguana, Jaragua and Salvatierra de la Sabana. His holdings were in the western part of the island, just across the Windward Channel from Cuba. Portly and pleasant-mannered, he was also ruthless; on Ovando's orders he had massacred friendly and peaceful natives of Española in wholesale fashion. His outward amiability also concealed a conspiratorial and insubordinate spirit. Cortés was to learn much from him.

Four ships carried the three-hundred-man expedition from Española to Cuba, and Cortés went along as secretary to Velázquez. He had participated in some of the skirmishes against the Indians in Española, and he was to see a little similar action in Cuba. But neither experience gave any hint of the military genius he was later to display. The natives were neither fierce nor warlike and their subjugation was easy. Only one Indian, named Hatuey, showed much resistance and spirit. He had fled from Española to Cuba after the massacre that Velázquez directed. There he was finally captured and condemned to be burned at the stake as a "rebel," although his attitude was more one of resistance than rebellion. A Franciscan friar urged him to be baptized as a Christian, since this would insure his entry into heaven. Hatuey is said to have asked if there would be Spaniards there. Assured that there would be, he declared that he wanted no part of it. Virtually the same story was later to be told of Moctezuma, the principal victim of Cortés's incursion into Mexico.

Cortés had begun to display traits that hinted at his future. "He was cunning and cautious," said one chronicler who was there. But, said his biographer, López de Gómara, "there was nothing to which he devoted more care than winning the good graces of his chief." He could not help note that Velázquez, while giving the appearance of fidelity and obedience to his chief, Diego Colón, did not hesitate to thwart and subvert the latter's orders and directions if (1) there was any likelihood of personal profit, and (2) he could get away with it. Velázquez was making every effort to switch his responsibility from Colón in Española to the Spanish crown. Before too long Cortés was behaving in the same way toward Velázquez.

Any understanding of the Cortés-Velázquez animosity, which was a key factor in much that happened later, is made difficult by contradictory, partisan versions and by the fact that it developed on two levels. On one level it was a power struggle — the kind that was a commonplace of Spanish public life, whether in Spain or in the colonies. On the other it was a complicated melodrama involving — once more — Cortés's relations with women.

The woman in this case was Catalina Xuárez (alternately Juárez and Suárez). The Xuárez family had come out to the Indies from Seville in 1509 in the same fleet that had brought Diego Colón and his vicereine, Doña María de Toledo. The Xuárez family consisted of Catalina, a beautiful and delicate young woman, her sister Leonor, their mother, known as La Marcayda, her brother Juan, his wife and their children. In the same party were Don Cristóbal de Cuéllar and his daughter, Doña María. Doña María was a lady-in-waiting to the vicereine, and the two Xuárez girls were ladies-in-waiting to Doña María. For two years or more the Xuárez family had endured hard times in Santo Domingo, the girls unable to catch the rich husbands they had hoped for. Doña María de Cuéllar, meanwhile, became engaged to Diego Velázquez. When Velázquez completed his easy conquest of Cuba Doña María went there, accompanied by her father, for the marriage ceremony to be performed. The Xuárez sisters, as ladies-in-waiting, went along, as did other members of the Xuárez family.

Unfortunately, Doña María died soon after the marriage ceremony. Velázquez consoled himself with the considerable and easily available charms of Leonor Xuárez. At the same time Cortés became involved with Catalina Xuárez — involved to the extent that Catalina expected him to marry her. This was an encumbrance that Cortés was either unready or unwilling to assume. Catalina was, in short, jilted. Her family and her friends did not take kindly to it. And the fact that her sister was the lover of the portly governor gave the private scandal a public air.

Cortés had, meanwhile, managed to get in trouble with the usually genial governor on another and much graver matter. Velázquez had given grants of Cuban land and Indians to some of his followers, using a system of preference and apportionment known only to himself. Cortés had been given some promising land on the Duaban River and a parcel of Indians to do his bidding — enough that he had no grounds for complaint. Others had received less than they considered their due, and some had received nothing at all and were understandably disgruntled. Some of the unhappy ones learned — perhaps through Cortés himself, who was in a position to know such things — that a panel of appellate judges had been sent out from Spain to Española to oversee the administration and execution of justice in the Indies. Secret meetings of the malcontents were held — and some of them were held in Cortés's house. Documents and evidence of Velázquez's unfairness were assembled and made ready for transmittal to the judges in Española. Cortés was the conspirators' choice as both spokesman and courier. Although it was a clear act of insubordination and involved a dangerous journey by native canoe across rough seas, Cortés agreed to do it.

Why Cortés, who had fared well at Velázquez's hands, should have taken part in such a plot is unclear. He may have resented Velázquez's intrusion into his difficulty with the Xuárez family. In the light of his later career it is easy to suspect that he detected a critical imbalance and a power vacuum and that, almost instinctively, he stepped in to provide needed leadership. Or it may have been that he loved legal argument and was eager to prove his skill at it.

Whatever the reason, word of the plot reached Velázquez. Cortés was arrested, fettered with chains and thrown in prison. Velázquez intended to hang him, but either his temper cooled or he was persuaded by Cortés's friends to relent. Then Cortés slipped his fetters, escaped from confinement and took sanctuary in the church. Here he might have been safe enough had not the wistful Catalina happened to show up in the churchyard. Gallant as ever, Cortés came out to talk to her. He was set upon by a squad of soldiers and personally seized by the commander of the squad, Juan Escudero (Cortés was to even this score some years later by ordering Escudero hanged in Mexico).

Once more in chains, Cortés was now put in the hold of a ship preparing to sail for Española to be tried by the same judges he had intended to appear before as an advocate. But he had no desire to appear before them as a prisoner. Once more he slipped his chains, stole a ship's boat and got away. The current, however, carried him farther and farther from shore. He shed his clothes and with the anti-Velázquez documents tied on his head in a handkerchief jumped in and managed to swim ashore. He again took sanctuary in the church, but not before he had a conversation with Catalina's brother, Juan Xuárez. It was a conversation in which Cortés must have promised to do right by Catalina. Juan Xuárez interceded and there was, finally, a reconciliation with Velázquez. It was a reconciliation that Velázquez would bitterly regret at a later date.

But for the time being the two men were reunited. Velázquez served as godfather for the baptism of Cortés's illegitimate daughter by an Indian woman. And a little later Cortés became formally betrothed to Catalina Xuárez, with Velázquez standing as his attendant. Catalina did not have an appropriate costume for the event and her brother, Juan, bought for her some clothing that had belonged to the late María de Cuéllar, Velázquez's short-lived bride.

The betrothal apparently had the effect of marriage. Cortés

lived part of the time with Catalina in Juan's house. He did not get around to providing a home for Catalina and a formal marriage ceremony until two years later, in 1515. By this time he was serving as municipal magistrate of Santiago and was prospering both through gold mining and livestock raising — with all the work being done, as usual, by Indians. His hacienda was said to be one of the finest in Cuba, and he was, at the age of thirty, a man of considerable substance and a fine-looking one too.

Bernal Díaz del Castillo, a young soldier of fortune from Medina del Campo, became acquainted with Cortés at about this time. He was, Bernal Díaz remembered later, "of good height and strongly made, with a somewhat pale complexion and serious expression. If his features lacked something it was because they were too small, his eyes mild and grave. His beard and hair were black and thin. His chest and shoulders were broad. His legs were bowed and he was an excellent horseman. . . . When he was angry the veins in his throat and forehead would stand out, and when he was very angry he would not talk at all."

Of the thousand or so Spaniards living in Cuba at the time, Cortés was one of the most affluent. His Indians found placer gold for him and the produce of the hacienda was profitable. He spent his money freely, gambled often, and while he avoided flashy dress, his clothes were quietly elegant, his jewelry good; his cloak was adorned with gold buttons in the form of knots. He was a generous host and the doors of his house in Santiago were open most of the time. He was one of the colony's most prominent men and he appeared, at first, to be perfectly content with things as they were. Most of his seven years in Cuba were as fat and prosperous as the seven in Española had been lean. He would have known of all the various expeditions that set out from Cuba, usually to capture slaves (the labor force was becoming critically short) and to discover new lands, but he was not tempted. Providing salt pork and bacon from his Duaban plantation for the outward-bound ships was a profitable business.

Early in 1517 Francisco Hernández de Córdoba made a disastrous visit to the "island" of Yucatán. The expedition, three ships and 110 men, suffered heavy casualties in clashes with the Indians there, and Córdoba died of his wounds not long after returning to Cuba. But he had brought back tales of temples and palaces of stone and mortar, walled cities, paved streets and proud, well-dressed natives who had a few gold trinkets.

Cortés may or may not have been intrigued, but Diego Velázquez was. He sent an agent to Spain to solicit a concession for future discoveries. And at the same time he mounted another and larger expedition, this one commanded by his nephew, Juan de Grijalva. Velázquez, apparently, trusted nobody, not even his nephew. When Grijalva had been gone longer than Velázquez thought reasonable, Velázquez began to suspect that Grijalva was betraying him, just as he, Velázquez, had betrayed Diego Colón. Perhaps Grijalva was claiming this new country for himself. To solve matters he sent another ship, commanded by Cristóbal de Olid, to search for Grijalva. And when Olid also was overdue to return he started planning still another expedition to search for the missing ones.

Velázquez's first choice to head the new expedition was Baltasar Bermúdez, a friend of long standing. But Bermúdez, just as demanding and ambitious as Velázquez, put so many conditions on his acceptance that the offer was withdrawn. Velázquez then turned to his former secretary and one-time enemy, Hernán Cortés. Cortés accepted.

Why, after showing no active interest in the earlier expeditions, was Cortés willing to undertake such a chancy and expensive undertaking? In retrospect it would appear that he had a sure instinct for the right thing at the right time. The Spanish had by now been in the Indies twenty-six years and they had little to show for it. They had found no quick and easy route to the Orient. Colonial administration had been an almost unbroken sequence of experiments and failures. A few fortunes had been made, but they were not large fortunes and had done little to enrich the mother country. The few placers and veins of gold — the Spaniards' greatest single preoccupation in the New World — were being exhausted. The native population was being steadily destroyed by harsh treatment, neglect and merciless exploitation. To a man of Cortés's acumen the need for change must have been apparent.

Cortés was given a long list of instructions. He was to search for Grijalva and Olid. He was also to search for castaway Spaniards who were thought to be living in that country of Yucatán, "the Rich Island." He was to keep a sharp lookout for Amazons, who were said to inhabit some of the islands (the supposed presence of Amazons gave both spice and motive to much of the Spanish exploration and discovery in the New World). He must treat the natives well and teach them about Christianity and the King of Spain. He and his men must not consort with Indian women (this would, if enforced, be a radical change for most of the Spaniards in the Indies). They were not to leave the coast to venture inland, were never to sleep on shore but must always return to their ships at night. They must never take anything from the Indians by force. All gold and other valuable goods must be kept in a locked chest. All lands discovered were to be taken possession of, but there was no mention of settlement to enforce such a claim. Velázquez obviously would attend to the settlement himself at a later date. He wanted to reduce the risk of insubordination and independence — the sort of insubordination and independence he had been guilty of.

Cortés assented to all of Velázquez's stipulations with good humor and an outward show of loyalty.

Meanwhile one of Grijalva's ships, commanded by Pedro de Alvarado, returned to Cuba with a cargo of some twenty thousand pesos' worth of gold obtained from the natives of this mysterious country. It was not a princely amount but it whetted interest in the Cortés expedition. Somewhat later Grijalva and Olid returned. Grijalva told of visiting an island called Cozumel, of cruising around Yucatán, of being received in friendly fashion in Tabasco, of sailing to the north and west and being visited by some finely dressed Indians who presented them with gifts of beautifully wrought golden jewels and ornaments. Although Velázquez had specifically forbidden Grijalva to attempt any settlement in this new country, he now upbraided him for not doing so.

Cortés was going ahead with his preparations. He began to dress even more elegantly than before and went about, people said, with a lordly air. He ordered banners made, embroidered in gold and displaying the cross, the royal arms and the legend: "Comrades, with true faith follow the sign of the Holy Cross and through it we shall conquer." This was the first public statement that the proposed voyage of exploration and trade was in reality to be one of conquest. More: he had a crier proclaim in the name of the crown, Governor Velázquez and himself — identified as the captain-general — that all who accompanied him would receive a share of the captured riches and a grant of land and Indians in the new country.

It was effective propaganda. Cortés soon had three hundred men signed up in the vicinity of Santiago. Spending freely, he began to accumulate supplies — pigs, chickens, bacon, beans, oil, wine, water kegs, barter goods such as glass beads, buckles, mirrors, scissors, ribbons, trinkets. But there were also stores that seemed suspiciously warlike: powder and lead, crossbows, muskets, ten bronze guns, four falconets and horses. Why, if members of the expedition were to live aboard ship and never venture inland, would horses be needed?

Velázquez's suspicions mounted. Cortés was gambling everything he had. He had spent something like twenty thousand pesos — as much as Alvarado had brought back in gold — on supplies, ships, horses. Much of the money was borrowed. Velázquez was to tell a friend: "I don't know what Cortés's intentions are toward me, but I think evil, for he spent everything he had on the expedition and is actually in debt, and received officials into his service with an air as if he had been a nobleman in Spain."

Anticipating a move by Velázquez to halt the expedition, Cortés finally made a hurried departure from Santiago on November 18, 1518, but not before raiding the local slaughterhouse and taking all available meat supplies. The little fleet sailed west along the south coast of Cuba, gathering additional ships, supplies and recruits on the way. At two stopping places local officials received orders from Velázquez to stop the fleet from proceeding, but Cortés, through a combination of flattery, diplomacy and bribery when necessary, managed to go on. All ships were to meet at Cape San Antonio at the western tip of Cuba. There, in February, Cortés held a muster. He had more than five hundred fighting men, a hundred or so sailors, altogether probably more than half the European population of Cuba. He had seventeen horses and an awesome arsenal.

He addressed his followers: "Certain it is, my friends and companions, that every good man of spirit desires and strives, by his own effort, to make himself the equal of the excellent men of his day. And so it is that I am embarking upon a great and beautiful enterprise, which will be famous in times to come, because I know in my heart that we shall take vast and wealthy lands, peoples such as have never been seen before, and kingdoms greater than those of our monarchs. . . . The lust for glory extends beyond this mortal life, and taking a whole world will hardly satisfy it, much less one or two kingdoms. . . . We are engaging in a just and good war which will bring us fame. Almighty God, in whose name and faith it will be waged, will give us victory. . . . We shall do as we shall see fit, and here I offer you great rewards, although they will be wrapped about with great hardships. . . . I shall make you in a very short time the richest of all men who have crossed the seas. . . . God has favored the Spanish nation . . . we have never lacked courage and strength and never shall. Go your way now content and happy, and make the outcome equal to the beginning." (Cortés reconstructed the speech many years later for his biographer, López de Gómara, and the words may well have been enriched by hindsight.)

And then the good men of spirit were off on the great and beautiful enterprise, to Yucatán or the Rich Island or whatever else it was that lay to the west.

THREE

Yucatán

CORTÉS, WHILE NEVER one to admit it, was at something of a disadvantage. Many of his followers were battlefield veterans. Some had even served under the Great Captain in Europe. Cortés was a comparative novice. At thirty-four he was junior to many of them. He knew nothing firsthand of the country whither he was bound. Thirty-eight of his men had been on the Grijalva expedition, and some of them thought Grijalva should be leading this expedition instead of Cortés. And a few of them had been on both the Grijalva and Córdoba expeditions.

To a man of less spirit this disadvantage might have been demeaning. But not to Cortés. Diplomacy and tact came to him naturally; he commanded his followers with an easy hand, flattering and cajoling when it was appropriate, bribing with promises if necessary. He was a gifted conversationalist, listened to everything, remembered much of it and put it to good use. Before reaching Cozumel, the island Grijalva had visited, he knew in detail the experience of the earlier expeditions and what they had observed of the country and the people.

The most experienced man of the lot was not a soldier but a pilot, Antonio de Alaminos of Palos. Alaminos had been a cabin boy on Columbus's fourth voyage, the one that had finally broken the Great Admiral's indomitable spirit. They had anchored in the lee of one of the Bay Islands off the coast of Honduras. While there, they were visited by a huge dugout canoe carrying an Indian trader, some twenty-five other men and a cargo of cotton goods and finery to trade with the island people. They were stronger and better dressed and appeared to be more intelligent than the Indians the Spaniards had known heretofore. They referred to their country as Maiam, or land of the Maya; and Alaminos remembered other words that were the same as words he heard later in Yucatán. An eager and curious youngster, Alaminos hoped that Columbus would go to the mainland and explore it. But the Great Admiral, almost insanely preoccupied with finding an ocean passage to Cathay, had sailed the other way, encountering nothing but storms, inhospitable natives and disappointments.

Just how the Yucatán peninsula, a squarish thumb of land thrust out into the Caribbean between 22° and 14° north latitude and 87° and 94° west longitude, so long escaped the notice of the eager Spaniards is one of the mysteries of the age of discovery. Several thousand miles of the New World's coastlines had been explored before Yucatán, so close to Cuba, was found. It may have been because much of the peninsula is a flat and almost treeless limestone plain, difficult to see from a distance.

Alaminos, by now an experienced pilot, served on both the Córdoba and Grijalva expeditions, as did both Bernal Díaz del Castillo and Francisco de Montejo. At Cape Catoche, on the tip of the peninsula, Córdoba's men could see in the distance a great city, which they called Grand Cairo. When they went ashore they were ambushed and driven back to their ships. The expedition's water kegs were leaky. It was an arid country and there seemed to be no rivers. Desperately searching for water, they went ashore at Champotón on the west side of the peninsula. Here they were attacked by a large and well-armed native army, which seemed completely unafraid of Spanish weapons and impervious to Spanish tactics. Nearly half of the invaders were killed and almost all the rest were wounded. Bernal Díaz suffered three arrow wounds, Córdoba ten. The survivors managed to sail away in their three ships and reach Florida, where Alaminos had been before with Ponce de León. Almost dying of thirst, they searched for water and again were attacked by warlike natives. Finally they struggled back to Cuba with little to show for their agony but they told many exciting stories about a formidable, tantalizing country.

By the time he joined the Grijalva expedition Alaminos knew something of the strong current that runs between Cuba and Yucatán — later to be known as the Gulf Stream. Navigation, while still not simple, was more purposeful. The expedition found and landed on the island of Cozumel, and later explored farther to the south along the coast of the "island" of Yucatán itself. Grijalva's chaplain recalled that ·they saw "three large towns about two miles apart. Many stone houses, very large towers and many straw houses could be seen in them. We would have liked to enter them if the captain had allowed us. . . . The following day near sunset we saw a town or village so large that the city of Seville could not look larger or better. In it a very large tower was visible [this was probably the ancient city of Zama and the modern archaeological site of Tulum, opposite the southern end of Cozumel]. Many Indians ran along the shore with banners which they raised and lowered, signaling us to approach them, but the captain was not willing. This day we came to a beach, close to which was a tower, the highest we had seen. A very large town was in sight."

The cautious Grijalva turned his ships north, rounded the peninsula and sailed along the coast of the Gulf of Mexico. At a bay they called Laguna de Terminos they went ashore, hunted deer and rabbits and sailed away, leaving a greyhound bitch behind. At the Tabasco River, which members of the expedition renamed the Grijalva in honor of their leader, they traded cheap green glass beads and brass bells to the friendly natives for objects of low-grade gold.

Grijalva and his men were able to converse with the natives after a fashion through the services of two Maya Indians, Julianillo and Melchorejo, that Córdoba had captured at Cape Catoche and brought back to Cuba. Both were cross-eyed (the Spaniards were to learn later than Mayan mothers often tied balls of beeswax to their infants' forelocks in the hope of getting their eyes to cross, a mark of distinction). Julianillo and Melchorejo were able to tell the Tabascans that the Spaniards wanted gold of better quality and more of it. To this the Tabascans replied with two strange words, "Colua" and "Mexico," accompanying the words with gestures toward the northwest. Later the Spaniards were to realize that both words referred to the Aztec empire, where most of the gold and other wealth of the country had been concentrated through conquest and tribute.

Grijalva sailed farther along the Gulf coast and anchored near the site of the present city of Veracruz. Here they were visited by natives far better dressed and of more stately demeanor than any they had encountered to date. These natives seemed eager to please the Spaniards and pressed on them gifts of rich costumes and golden jewels. This provided the bulk of the treasure cargo that Grijalva dispatched back to Cuba in the ship commanded by Pedro de Alvarado. With the remaining three ships Grijalva cruised a little farther north before retracing his route and returning to Cuba.

From accounts of the earlier voyages Cortés was able to put together a sketchy picture of the country and its people.

In many places the warriors wore quilted cotton armor which could stop an arrow or a spear. Cortés ordered his men to equip themselves similarly; the cotton was lighter and cooler than their customary iron armor. The natives were skillful archers (although they did not poison their arrows as did many Indians elsewhere). They also used spears and heavy two-handed wooden swords edged with razor-sharp inlays of volcanic glass. From men they killed they sometimes cut away the jawbone and wore it as a trophy on the upper arm. War prisoners and others were sacrificed at their temples and the natives ate the flesh of the victims. They worshipped hideous idols. And they were said to indulge in sodomy.

At the encounter with Córdoba's men the natives had shouted something that sounded like "Castilan, Castilan!" It seemed to suggest that the rumor of castaway Spaniards, or Castilians, in Yucatán was true.

The peninsula was for the most part an arid, calcareous plain. Most of the water came from sinkholes in the limestone, and the natives guarded their water supply jealously. But despite this scarcity the natives bathed frequently. To the Spaniards this suggested the hated Moors; they, too, were frequent bathers. The Spaniards tended to think too much bathing sapped a man's vitality.

To the south and west of the peninsula were mountains, jungles and rivers. From the anchorage where Grijalva had received the rich gifts, a place he called San Juan de Ulúa, one could see many high ranges and one snow-capped peak. Beyond those mountains someplace lay the mysterious Colua and Mexico. This must be the source of the gold Grijalva had received at San Juan de Ulúa.

Pedro de Alvarado, who had been the courier for Grijalva's gold, was one of Cortés's captains. A redheaded, impetuous, high-spirited man, he was put in charge of the *San Sebastián,* one of the eleven ships of the Cortés expedition, with a man named Camacho as pilot. Alvarado and Camacho ignored Cortés's order for all ships to rendezvous at Cape San Antonio. Instead they sailed ahead and arrived at Cozumel two days before the rest. By the time the ten other ships arrived they had explored two abandoned towns, stolen forty fowl and looted the temples of some insignificant trinkets of low-grade gold.

When Cortés arrived he was furious. He ordered Camacho thrown in chains, and he publicly dressed down the venturesome Alvarado (it was to be only the first of many times that Alvarado would embarrass his leader). Through a woman prisoner who spoke the dialect of Jamaica and Cuba Cortés sent an apology to the natives who were in hiding and invited them to the Spaniards' camp. When they were finally coaxed in, Cortés returned the objects stolen from the temples. In payment for the fowl, which had been eaten, he gave them gifts of bells, scissors, knives and the always useful green beads (the Spaniards quickly discovered that the natives identified the cheap beads with their own *chalchihuitl,* or jadeite, which they considered more precious than gold).

Cortés had with him Melchorejo, the Indian whom Córdoba had brought back and who had sailed with Grijalva (Julianillo had since died). Melchorejo's comprehension of Spanish and his loyalty to the Spaniards were both questionable. Nevertheless Cortés managed, through him, to persuade the peaceful and friendly natives of Cozumel to allow him to break up the idols in their temples and to erect crosses and a shrine to the Virgin. It was one of the easiest and most effective conversions of the entire conquest. For years afterward voyagers reaching Cozumel would be greeted with cries of "María! María! — Cortés! Cortés!"

Cortés also inquired about the castaway Spaniards. He was told that several Spaniards who had been in a shipwreck were held captive on the mainland, across the Cozumel channel. The Cozumel people were afraid of the more warlike mainland Indians. But finally one man was persuaded to serve as a messenger and was taken aboard a ship to Cape Catoche. He was given a quantity of the magic green glass beads and a message signed by Cortés: "Noble lords, I departed from Cuba with a fleet of 11 vessels and 550 Spaniards, and arrived here at Cozumel where I am writing this letter. . . . The people of this island have assured me that in your country there are five or six bearded men like us in every respect. . . . I suspect and consider as certain that you are Spaniards. I beg you within six days from the time you receive this letter, to come to us without delay or excuse. . . . We shall recognize and reward the favor. . . . I am sending a brigantine to pick you up."

The courier hid the message in his rolled-up hair, was deposited on the mainland shore and disappeared. The brigantine waited for eight days. On the ninth the crew, concluding that the messenger had either run away or been captured by hostiles, sailed their craft back to Cozumel.

But the courier had done his work. He had found one of the missing Spaniards, Jerónimo de Aguilar, who was held as a slave by a *cacique,* or chief. Aguilar gave some of the green glass beads to his *cacique* and was freed.

Aguilar had been in Yucatán eight years. In 1511 he had sailed from Darien in the Panama isthmus for Española. The ship on which he traveled, commanded by Juan de Valdivia, struck some shoals called Las Víboras off the coast of Jamaica and had broken up. Nineteen survivors managed to climb into one of the ship's boats and drift away. Some died at sea of hunger and thirst. Those who were left came ashore on the coast of Yucatán and almost immediately were taken prisoner by the natives. Some, including Valdivia, were sacrificed and eaten. The others, emaciated from their ordeal, were placed in wooden cages to be fattened up for future sacrificial feasts. But they escaped from their cages and scattered. Aguilar fell into the hands of a *cacique* who had no interest in sacrificing him. But he did have other ideas. He presented Aguilar with an enticing young woman, perhaps with the idea of producing a new blood strain in his subjects. But Aguilar, who had taken religious orders in Spain, stoutly resisted the young woman's charms and was thereafter put to work as a menial.

There was one other survivor, Gonzalo Guerrero. Guerrero had fallen into the hands of still another *cacique.* He attracted the attention of the *cacique*'s daughter, who insisted that he be hers. Guerrero had none of Aguilar's inhibitions and was eager to please. They married in the native fashion and had three children. Although he still sometimes continued to think of himself as a Spaniard, Guerrero had adopted native ways. He was tattooed, wore ear pendants and had his lower lip fitted with a decorative stone plug. He taught his adopted brethren European modes of fighting and warfare, how to flank an enemy instead of charging head-on in the Indian fashion, how to build and utilize bastions and strong points. His coaching of the Indian troops had proved itself most effective in the battle with Córdoba's troops at Champotón. He was, in short, a thoroughly integrated Indian and a highly respected one.

Aguilar went to Guerrero, showed him Cortés's message and urged him to go with him to join his countrymen. Guerrero sadly told Aguilar that it was too late for him to do so. But he asked that Aguilar leave some of the green glass beads so that he could tell his sons they were gifts from his native land. Meanwhile Guerrero's wife upbraided Aguilar, a lowly slave, for even daring to approach her warrior husband.

By canoe Aguilar set out for Cozumel. He arrived just as the flotilla was sailing away, the leader having given up all hope of finding (1) gold or (2) missing Spaniards. But before they were far out at sea, the ship commanded by Juan de Escalante, heavily loaded with cassava bread, the expedition's principal staple, sprang a leak and all ships were ordered back to Cozumel. While waiting for the ship's seams to be caulked, some of the men were hunting wild pigs. On the beach they met a man whom they at first thought to be an Indian. He was nearly naked, burned dark by the sun. He wore one sandal and carried another. He knelt on the sand in native fashion, gave thanks to God in broken Spanish and asked somewhat incoherently if they were Spaniards. Assured that they were, Aguilar identified himself and asked what day it was. From a scrap of cloth he unfolded a tattered Book of Hours in which, for eight years, he had tried to keep track of the days. His calculation was that this was Wednesday, and since he always said extra prayers on Wednesday he knelt again and asked the Spaniards to join him. Actually it was Sunday.

Cortés gave Aguilar clothing and questioned him closely about the mainland, the number of natives, their military strength. To most of it Aguilar replied that he had been nothing but a slave and knew very little except hewing wood and tilling the earth. Cortés was interested in the story of Guerrero and his instruction of the natives in military tactics. "I wish I had my hands on him," he exclaimed, "for it will never do to leave him here." But he did nothing further about it.

What was more important to him at the moment was that he now had in his company a man who was not only fluent in the Indians' language but was also at home in Spanish, the abstruse dogma of the Catholic church and the Spanish crown. Melchorejo was no longer the expedition's "tongue," and not long afterward he hung his European clothes on a tree limb and disappeared forever.

Neither Aguilar nor anyone else could tell the Spanish adventurers much of the strange past of this country, nor did Cortés display any curiosity about it. The Mayan realm had stretched from the Yucatán peninsula deep into Central America and southern Mexico. The Mayans had begun to achieve greatness before the Christian era and had become the most advanced people of ancient America. They had made astonishing discoveries in astronomy and mathematics, and had even invented the abstract concept of zero to facilitate their calculations. They had devised a calendar more efficient and accurate than anything known to the Europeans. And they had studded their plains and jungles and mountains with magnificent temples and palaces. But the Mayans' empire had begun a long and steady decline more than six centuries before the Spaniards arrived. The Spaniards, greatly attached to their own antiquities, cared less for those of other peoples, and they saw no reason to consider the Mayans as anything more than benighted heathens who had very little gold.

Cortés and his men were eager to be off in the direction of the country where there was said to be much gold. Weapons were cleaned and oiled. Swords and lances were sharpened.

Spare parts were made for the crossbows. There was target practice and drill for the troops. Many of them cut off their cherished queues since they might be a handicap in hand-to-hand fighting. And finally they departed. They carried a large supply of honey and beeswax provided by the Cozumel people. Becalmed, they dug salt from the evaporative flats at Isla Mujeres. A giant shark was captured and cut open. In its stomach were found ten sides of bacon that had been hung over the side of one of the ships for soaking in the salt water. They avoided Champotón, where the fierce and well-trained Indians had inflicted disaster on Córdoba. During a storm one of the ships disappeared. It was found later moored in a bay. The greyhound bitch left behind by Grijalva's men greeted them from the beach. The dog was fat and sleek and she dashed away to bring in a plentiful supply of rabbits. The missing ship's rigging was hung with rabbit meat, drying in the sun.

Thus far, except for minor mishaps, it had been like an excursion, an outing. Finally the ships reached the mouth of the Tabasco, or Grijalva, River. Knowing of Grijalva's experience there, Cortés expected a friendly reception and, perhaps, more gold than he had found at Cozumel.

FOUR

Tabasco

JUST AFTER COLUMBUS'S first voyage of discovery, Pope Alexander VI had ordered that natives of newfound lands be instructed in Catholicism in the interest of the "salvation of these souls . . . so that barbarous nations may be humbled and converted to the Faith." Subsequently Spain's Council of the Indies drafted a *requerimiento,* or notification, that conquistadors were expected to read and explain to all natives encountered in Spain's imperial expansion.

Reduced to its simplest terms the *requerimiento* said that God had, long ago, created the first man and the first woman. Their offspring scattered in many lands around the world and multiplied. Although they might not know it, they were all God's creatures. God had designated St. Peter as their chief. St. Peter's successor, the Pope in Rome, had given the islands and mainland of the Ocean Sea to Spain. Therefore those of God's children who were in these remote lands were subjects of the Spanish throne, whose vassals they must acknowledge themselves to be. If this subjugation to God and vassalage to the Spanish throne were accepted, well and good; the subjects could expect favors, honors and benign treatment and would be made privy to the wonderful secrets of Christianity. But if they rejected it they were clearly in rebellion and would suffer war, punishment, slavery.

Many conquistadors did not bother with the *requerimiento.* They could barely understand it themselves; how could they explain it to ignorant savages? But Cortés, who was to be responsible for snuffing out perhaps a quarter of a million lives in the New World, was punctilious about it. It had been unnecessary to employ it at Cozumel, where the natives were tractable and seemingly happy to accept both Christianity and vassalage. But it was to be different at Tabasco.

The little fleet sailed west along the coast. Rivers, almost nonexistent on the Yucatán peninsula, began to appear more frequently. Between the rivers were swamps and broad alluvial plains. Finally they came to the Grijalva, where the natives were expected to be friendly. A sandbar blocked the mouth of the river. The large ships had to be anchored outside. Cortés and eighty of his men boarded a brigantine and some of the ships' boats and proceeded up the river, carrying guns. A little more than a mile upstream they came to a settlement on the west bank, a dense cluster of adobe buildings with palm thatch roofs surrounded by a log stockade. Canoes loaded with natives, most of them armed with bows, came out to meet the strangers.

Through Aguilar Cortés said that he came in peace and that he wanted only to buy supplies of food and water. In fact, neither was needed. His ships still had an ample supply of staples brought from Cuba, and the river was full of fresh water. But it served to open the conversation.

The Indians left and soon returned with several canoeloads of food. This was not enough, Cortés told them, and asked permission to enter the town to look for more. The Indians said they would make a decision the next day. Cortés and his men camped on an island in the river. During the night the Indians evacuated women and children from the town, preparing for the worst. Cortés meanwhile sent orders for all the fighting men on the ships to join him. When he had been reinforced, he sent several hundred Spaniards upriver to a point where they could cross; they hid themselves in the woods in position to attack the town from the rear.

In the morning eight canoes approached Cortés and his men on the islands. The Indians had brought a small additional supply of fruit, fish, fowl and tortillas, thin cakes made from ground maize. This was all there was, they said, because the town was deserted. Cortés replied that this was not enough and proposed to enter the town. The natives protested that he had no right to do so, whereupon Cortés ordered Aguilar to read the *requerimiento,* the prerequisite for forcible invasion. It had no effect on the Indians. They would oppose any entry by the Spaniards and again asked that they depart in peace. Cortés replied that with God's help he would sleep in the town that night. The Indians laughed and left.

The Spaniards ate, cleaned their weapons and prepared to fight. Cortés clasped his shield and dramatically called on the patron of Spain, St. James, and his personal patron, St. Peter, to support them. He sent word to the soldiers in the woods and then, at the head of two hundred men, approached the village.

The Spaniards had to step from the boats into thigh-deep water. Cortés lost a shoe in the bottom mud. Meanwhile a hail of arrows, darts and stones from slings poured on them from apertures in the log stockade, while the Indians shouted, screamed and blew on conch shell trumpets. Twenty Spaniards were wounded. The town was being attacked from both the river in front and from the woods in the rear. Soon the attackers were inside the stockade and after a few minutes the only Indians left were either dead or wounded, and a few survivors were made prisoners. Cortés set up his quarters in the stone temple and ordered a search of the houses. Some stocks of maize and some cotton garments were found, but no gold. Cortés with his sword cut three slashes in the trunk of a giant silk-cotton tree and claimed the country for Spain.

On the morning of the third day Cortés sent word by the prisoners that he wished to meet the *cacique,* or chief, to explain that bloodshed had been unnecessary and was the Indians' fault. The invitation was ignored. Cortés then dispatched three scouting parties to the maize fields beyond the town. There were several skirmishes, and a few more prisoners were taken. The prisoners informed the scouts that Indian troops were being mobilized in the entire region for a mass attack on the Spaniards, and that all of the invaders would be killed and eaten.

But for several days little happened. A party of twenty Indians came to Cortés with a request from their *cacique* that the town be spared, that it not be burned. They brought with them a supply of fruit. They refused to accept pay for it, and Cortés again insisted that more food was needed.

Scouting parties again became involved in a fierce battle with the Indians in a village near the maize fields. Cortés himself went to their relief with a company of one hundred men and some artillery pieces. It was the first time artillery fire had been used and it created havoc in the Indian ranks.

That night, the sixth on shore, Cortés prowled his camp all night long, talking to his men and inspecting equipment. The wounded were loaded into boats and taken to the ships, and the horses brought back, ready for what would be the first cavalry action on the mainland of the New World.

In the morning, after mass, Cortés assembled his foot soldiers, artillery and cavalry — thirteen horses were ready for use — and set out for the nearby plain of Ceutla, where Indians from all the region had assembled — five squadrons of eight thousand men each. The plain was cross-hatched with drainage canals, and the earth had been cultivated. It made for uncertain footing for the heavily armed Spaniards. Cortés deployed his horsemen in a wood where the Indians could not see them. Meanwhile the foot soldiers advanced across the plain toward the massed Indians, often stumbling and falling in the ditches and loose earth.

Bernal Díaz, who was fighting on foot, described the scene: "As they approached us their squadrons were so numerous that they covered the whole plain, and they rushed on us like mad dogs, completely surrounding us, and they let fly such a cloud of arrows, javelins and stones that on the first assault they wounded 70 of us, and fighting hand to hand they did us great damage with their lances, and one soldier fell dead at once from an arrow wound in the ear and they kept on shooting and wounding us. With our muskets and crossbows and good swordplay we did not fail as stout fighters, and when they came to feel the edge of our swords little by little they fell back, but it was only so as to shoot at us with greater safety. Mesa, our artilleryman, killed many of them with his cannon, for they were formed in great squadrons and they did not open out, so that he could fire at them as he pleased, but with all the hurts and wounds which we gave them we could not drive them off."

The Indians shouted, whistled, blew on their conch shell trumpets and cried "Alala, Alala!" — according to Bernal Díaz — and threw dust and grass in the air to hide their losses. They were taking heavy losses but were not weakening.

The Spanish foot soldiers had begun to wonder why Cortés and his horsemen did not come to their aid. The little cavalry troop was trying to do so, but the ditches and swampy ground had delayed them. However, one horseman did appear, mounted on a dapple gray. He charged the massed Indians again and again, slashing with his sword. This was the first glimpse the Indians had had of a man on horseback; they mistook the two of them for one monstrous centaurlike creature and were terrified.

Many of the Spaniards insisted that the appearance of the rider of the dapple gray who had arrived in a time of peril was a miracle, that it was surely St. James, for whose intercession many of them had silently prayed.

When Cortés himself arrived on the scene a few minutes later and was told of what had happened, he said he thought it probably was his own patron, St. Peter, who had intervened. And he shouted to his men, "Forward, companions — God and glorious St. Peter are with us!" Bernal Díaz did not subscribe to the miracle theory; he had seen no mysterious horseman, but he conceded that perhaps he had been unable to do so because he was a sinner. Others, more realistic, insisted that the lone horseman was Francisco de Morla, who had simply advanced more rapidly than the rest of the cavalry.

When all thirteen horses were in action, the Indians melted away before them and the outcome was quickly settled. Three hundred Indians lay dead on the field and many more had been wounded by gunshot, crossbow bolts and slashing swords.

Within two days the Indians politely asked permission to gather and bury their dead. Then they treated for peace, bringing supplies of food to the Spanish camp and gifts of roses and turkeys for the horses, of which they were still terrified. Cortés intensified their terror with a ruse. A mare that had recently foaled was concealed in the camp. Then, with the Indians watching, he led in a stallion — "a lewd stallion," said one of the chroniclers. At the scent of the mare the stallion trumpeted, pawed the earth and repeatedly reared on its hind legs. Cortés explained that the horse was still angry with the Indians. Then he spoke gently to it and had it led away, to the Indians' relief.

The Indians had also brought what gold they had, a few hundred pesos' worth, and some bits of turquoise. It was pitifully little by the standards of the Spanish hopes and dreams. To Cortés's questions they replied that they did not know what gold mines were, that they picked up gold when they happened to encounter it but that they did not go looking for it, since it meant very little to them. And again there was mention of the powerful country toward the sunset, where the rulers coveted and collected gold.

Cortés had other questions. Why had the Tabascans fought him when, the year before, they had welcomed another Spanish captain? The natives answered that the other Spaniards seemed only to want to trade for gold, and in this they had accommodated them. When Cortés's much larger and more awesome fleet had arrived, they decided that the strangers had come to take by force whatever was left. It was also suggested that the warlike natives of Champotón, who had inflicted such punishment on Córdoba's men, had ridiculed the Tabascans for not waging war on Grijalva's expedition.

All in all, Cortés's operation in Tabasco seemed in retrospect both meaningless and futile. Cortés later authorized a dispatch to Spain describing his motives: "In his devotion to the services of your Majesties and being desirous of sending a true report of all there is to know in this new land, [Cortés] decided to proceed no farther along the coast until he had discovered the secret of the river and of the towns which are on its banks, for they are rumored to be fabulously rich in gold."

The rumored riches most likely came from Cortés's versatile imagination. The Grijalva expedition had obtained very little gold in Tabasco, and no other Europeans had been there.

If Cortés had been intent on establishing a colony, a bridgehead on the mainland, Tabasco would have been a good place for it. It had fertile land, great rivers, forests of hardwood, an abundance of game and native fruit, a large and healthy Indian population. Yet he made no effort to settle and colonize. Nor did he really explore the surrounding country — if he had he might have discovered the spectacular Mayan ruins of Palenque not far away, mute evidence of the skills and culture these people had once had. He did not need the food and water he demanded; he used it only as an excuse for waging a little war in which perhaps as many as eight hundred Indians were slain, a matter that Cortés later described as "the will of God." The Dominican friar Bartolomé de las Casas said that Cortés entered the country "as a tyrant" with his "holy company." Later there was to be a royal investigation into Cortés's actions in Tabasco.

In retrospect Cortés's motives seem simple. For one thing he had not, before Tabasco, proven his own capacities as a battle-field commander. He needed to demonstrate his courage, his skill with arms, his qualities as a leader, both to his own satisfaction and to convince his little army that he was a man worthy of being followed and obeyed. He also needed to observe Indians, how they fought, what punishment they could take and, particularly, their reaction to gunpowder, Toledo steel and men on horseback. As such things went it was a cheap experience. One of his men had died and several score were wounded, none gravely. And he had gained much more than he at first realized.

April 17 was Palm Sunday. Cortés ordered his men to erect an altar and a wooden cross. He insisted that the natives come to see the strange sight of these fierce white men carrying palm fronds instead of swords and prostrating themselves meekly before the cross while a priest said prayers and retold the miracle of the Prince of Peace.

Another part of the ceremony was the christening of twenty young women the vanquished Tabascans had presented to the Spaniards as a gift, to grind their maize, mend their clothes and provide whatever other services and comforts might be required. Cortés decided the women should be distributed among the more important of his followers as *camaradas,* a euphemism for what would obviously be the dual role of servant-concubine. It was forbidden that a Spaniard should lie with a native woman who had not been christened. In the past, religious rectification often came long after the carnal act if at all. And Cortés himself had observed no such stricture during his residence in Española and Cuba, although in Cuba, after having a child by an Indian woman, he had seen to it that she was baptized. But in Tabasco, with the expedition on a war footing and with discipline very tight, the christening was performed promptly.

Most of the women were short and plump, seemingly bone-less. But one was much taller and had a regal bearing. It was clear that she was a person of some importance. In the twenty years of life that remained to her she was to establish herself as a towering figure in the destiny of Mexico, second only to Cortés himself. The Spaniards sensed that this was no ordinary Indian woman. She was christened Marina and given the courtesy title of *doña* — remarkable in that not one of the five-hundred-odd conquistadors could at that time claim to be a *don*.

Marina was, at the time, somewhere between fourteen and nineteen years old. Her life both before and after her presentation to the Spaniards has engaged the attention of generations of historians, poets, novelists, philosophers, philologists — even astrologers. While there are some reliable data on her subsequent career, her early life is clouded in mystery and myth. There is some evidence that she came from the province of Coatzacoalcos, adjacent to Tabasco in what is now the southeastern part of the state of Veracruz; and that her home-town was probably Oluta. The area is part of the Isthmus of Tehuantepec, a region that for centuries has been famous for its women: stately, beautiful, invariably strong-minded and, often, remarkably relaxed in both manners and morals.

Her name originally was believed to have been Malinali Tenepal. The first name is the same as the twelfth day of the twenty-day Aztec month, probably the day of her birth. The last name is believed to have connoted a person who speaks much and with animation. The area was under Aztec domination and the language was Nahuatl, unlike the Mayan dialect spoken in Tabasco.

Malinali was said to be the daughter of a *cacique* who died while she was still young. Her mother, Cimatl, soon remarried, to another *cacique*. A son was born and the mother feared that Malinali might stand in the way of her half-brother inheriting the chieftainship. Accordingly she sold the girl to some traveling traders, who later resold her as a slave in Tabasco. To conceal the cruel act an even crueler one was committed. The body of a slave's daughter, deliberately killed for the purpose, was displayed as proof that Malinali had died.

The slave girl's christened name, Marina, may have been a Spanish approximation of Malinali. But to the natives Malinali-Marina became Malinche, the suffix *-che* indicating respect.

But Cortés was not Malinche's lord immediately. In the distribution of the slave girls he presented her to Alonso Hernández de Puertocarrero, one of his captains whom he particularly favored whenever possible.

On the Monday after Palm Sunday the fleet sailed away from Tabasco. And on Good Friday, the same day on which in 1504 Cortés had had his first glimpse of the Indies, the ships anchored at approximately the present site of the city of Veracruz, the place that Grijalva had called San Juan de Ulúa.

Here Cortés and other Spaniards discovered something else about Marina. She was overheard talking with some natives who came to the shore to see the strange sight, but speaking in a different tongue. It was Nahuatl, the language of the Aztecs and their tributaries.

With this bilingualism she could translate from Nahuatl into Mayan, which Aguilar could understand and translate into Spanish — and before very long Marina had learned enough Spanish that she could translate directly. But she was to be much more than interpreter. She could not only tell Cortés what the words meant but could also explain Indian attitudes, expressions, gestures, acts and reactions. She was sensitive to everything that went on, an acute observer. Her quick wit would later save Cortés and his men from entrapment and slaughter.

Puertocarrero was soon sent off on a mission to Spain. Marina thereafter rode behind Cortés on his horse, stood beside him in the field, shared his bed at night and later bore him a son.

FIVE

Tenochtitlan (I)

CORTÉS HAD, THUS far, been the beneficiary of two phenomenal combinations of luck and happenstance.

First: the recovery of Aguilar and his knowledge of the Mayan language and, because of his resistance to the charms of an alluring native girl, his willingness — unlike Gonzalo Guerrero's — to serve his countrymen.

Second: the acquisition of Doña Marina or Malinche, with her knowledge of both Mayan and Nahuatl languages and her well-founded reasons for hating the Aztecs.

There was to be a third element of fortune that Cortés could not have planned, hoped for or even dreamed. It was that he would be mistaken for a god. The mistake was rooted in the dark and complex history of the people he was to conquer. It was tangled in the antipathetic characters of two old gods with tongue-twisting names and strange powers. And it was nourished by the tortured indecision of a native ruler whose fate, finally, lay in Cortés's hands.

This was Moctezuma Xocoyotzin, the Angry Young Lord, otherwise known as Moctezuma II, ninth ruler of the Aztec empire, the inland nation that the Tabascans had identified as

"Colua" and "Mexico," the place where all the gold was. Moctezuma had occupied the throne since 1503, the year before Cortés set out for the Indies.

The Aztecs, or Mexica, had only recently risen to power. The empire or confederation that they dominated was little more than a century old. But although comparatively new it was impressive, stretching from northern Mexico to Guatemala. It was also tyrannical and autocratic.

Older and grander civilizations had risen, reached peaks of culture and affluence and declined before the Aztecs arrived on the scene. These older civilizations were quite as baffling to the Aztecs as the Aztecs were to be to the Europeans. In central Mexico, where Moctezuma's people reached the zenith of their power, there had been at least two great cultures long before the Aztecs emerged as a nation. The sacred city of Teotihuacan, with its gigantic pyramids, had flourished between the second century B.C. and A.D. 750. Then it had lapsed into desolation and solitude. The Toltec civilization of Tula, or Tollan, had risen as that of Teotihuacan had collapsed. By A.D. 1200 it too had begun to decline.

One of the reasons for the decline was the number of comparatively uncivilized people who drifted in from the north and west, much as the Germanic hordes had descended on Rome. There were at least seven and possibly as many as eleven tribes of these migrant peoples who were to assume power in central Mexico. One of the latest to come was the Aztecs, the people of Aztlan, or the Place of the Herons.

The location of Aztlan was then and remains a mystery. It was supposed to be an island, rich in fish and bird life, surrounded by rushes. It was also believed to contain some caves, in which the ancestors of the Aztecs dwelt. But more precise information had been forgotten; or it had been deliberately distorted by one Aztec ruler who, in the apparent hope of upgrading the tribe's humble origins, had ordered tribal history revised and made more distinguished.

Moctezuma's grandfather, Moctezuma I, perhaps the

greatest of the Aztec emperors, was curious about his people's past. He selected sixty sorcerers and sent them off to find Aztlan. They retraced part of the tribe's migratory path. Then, they said, through spells and incantations they were transformed into birds and beasts for the rest of the journey. They had a conference with an incredibly old and dirty goddess named Coatlicue, She of the Snaky Skirt, who was the mother of their own tutelary god, Huitzilopochtli. She presented them with a maguey fiber mantle and a loincloth to deliver to her son. Then the sorcerers reverted to human form and returned to report to Moctezuma I, unable to provide any information about the Aztecs' origins or where they had been.

From more reliable sources it has been deduced that the Aztecs left their original home, wherever it was, early in the twelfth century and spent more than one hundred years wandering, living sometimes as nomads, sometimes as settled agriculturalists. With them they carried the idol of Huitzilopochtli, always in the custody of four priests who interpreted the god's directions to his people, where they should go, how long they should stay, where they should settle permanently. This final settlement was on a marshy island in the great lake of the valley of Mexico. Here the Aztecs — who, on the god's instructions, had changed their name from Aztec to Mexica — found the promised sign: an eagle perched on a nopal cactus devouring a serpent. It was a poor location but it was enough for the hardy but travel-weary Aztecs, or Mexica. Here they built their city, Tenochtitlan. By the time the Spaniards saw it first in 1519 it was the equal of any city in Europe in population and grandeur — an island city laced with canals and paved thoroughfares, a place of great pyramidal temples and luxurious palaces, well-built dikes to prevent flooding, an aqueduct to bring sweet water from the heights of Chapultepec, straight causeways linking the island with the shores of the lake, great botanical and zoological gardens, vast markets in which treasures and foodstuffs and human slaves from all parts of the realm were available in staggering abundance.

The principal reason for the Aztecs' astonishing growth from marginal nomadism to imperial might was their military skill and their insatiable appetite for conquest. When they had arrived in the valley, lowly and scorned by the longer-established tribes, they had been forced into other peoples' wars. But as hired soldiers they demonstrated both courage and ruthlessness. Soon they were respected and feared in their own right. They began conquests of their own against other city-states. With those they could not conquer they formed alliances.

Still, they suffered from a sort of inferiority complex. When they decided to choose a *tlatoani,* or king, they had no suitably impressive candidates among their own people. They went to the neighboring and longer-established city of Culhuacan and asked for the services of a satisfactorily aristocratic nobleman to come and rule them. The one chosen, Acamapichtli, took the throne in 1372. With a considerable harem of wives and concubines he became the progenitor of most of the Aztec royalty and nobility. Sharp distinctions were made between royalty, the nobility, ordinary folk and the slaves. The reconsideration and revision of Aztec history was also undertaken on his orders. One of the results was an elevation in the terrifying importance of Huitzilopochtli, who came to be recognized as the god of war, the god of the sun and almost everything else.

Originally perhaps Huitzilopochtli had been an ordinary god in the Aztecs' dim past, or he may even have been a human, a greatly revered tribal leader. During their migration the Aztecs acquired other gods to add to their pantheon, but none of the others ever achieved the awesome powers of Huitzilopochtli. As the problems of statehood grew, so did Huitzilopochtli's appetite for human sacrificial victims. Many primitive religions go through a phase of human sacrifice and ceremonial cannibalism, but few if any ever reach the heights of rapacity exercised by the Aztecs in the name of their principal god — whose name, disarmingly, meant Hummingbird of the South; the figure usually displayed a long, sharp, golden bill, like that of a giant hummingbird.

Captives taken in small wars were sacrificed as a matter of course. But larger national needs and ambitions demanded more favors of the god. The result: larger wars, more captives, more sacrifices. Defeated states were made to pay tribute in gold, jewels, clothing, foodstuffs and, most important, victims for the sacrificial stone. When military captives and human tributes failed to provide enough victims the Aztecs engaged in "Flower Wars" with friendly or semifriendly tribes. These were completely artificial encounters, held at prearranged times and places, with no purpose other than to provide an opportunity for the taking of prisoners: that is, sacrificial victims.

In 1487 a great new temple to Huitzilopochtli was completed in the heart of Tenochtitlan and dedicated in the greatest sacrificial bloodbath in Aztec history. Invitations were sent out to subject peoples and to independent states, both friendly and unfriendly. Each was to send sacrificial victims. To refuse was to invite attack by the Aztecs. Ahuizotl, an uncle of Moctezuma II, had only recently ascended the throne. In preparation for the event he had waged war along the Pacific coast and collected a stupendous number of prisoners.

The stucco surfaces of the stone temples were freshly painted in brilliant colors for the celebration. The doomed captives, decorated with paint and feathers, formed interminable queues along the causeways leading into the city. The temples were decorated with flowers and the air throbbed with the sound of wooden drums. Ahuizotl and the allied rulers of Texcoco and Tacuba, each dressed as an Aztec deity, had a substantial lunch. Then the three kings ascended the steps of the main temple, stretched their victims one by one on the sacrificial stone and, using flint knives, ripped the hearts out of their chests. Blood from the hearts was then sprinkled around the temple and on the idol. The air was thick with the reek of blood. The heartless bodies were then rolled down the steps past the thousands more victims who were awaiting their turn. Later the bodies were cut up and distributed for ceremonial feasting. Chroniclers reported that eighty thousand were thus sacrificed on a nonstop basis during the four-day celebration — although the figure, more than one-fourth the total population of Tenochtitlan, seems unlikely. On the fifth day the guests, friends and enemies alike, departed, laden with gifts from Ahuizotl.

It was not always battlefield captives or token captives that were sacrificed. The death of a nobleman was usually marked by the funeral sacrifice of dozens or even hundreds of lesser beings, selected, of course, from among the plebeian classes. Although there was always a certain false tenderness attending such sacrifices — the killer addressing the victim as "my son" (or daughter) and the victim addressing his or her executioner as "my father" — the business of ceremonial murder built up resentment among the survivors. Such resentment was aggravated by a harsh social order. Capital punishment was inflicted on the lower classes not only for murder but also for such lesser crimes as adultery — which was practiced freely by the aristocracy — homosexuality and second-offense drunkenness. It could also be imposed if a common man looked in the face of a nobleman or overtook him while strolling. He could also be killed for wearing a garment above his station, for entering a nobleman's house, or for felling a tree for fuel. If accused by any nobleman of committing treason he was tortured to death, his house torn down, his fields strewn with salt and his family and descendants enslaved for four generations.

By the time Moctezuma II assumed the imperial throne these resentments had begun to reach alarming proportions. Also there had been crop failures and reverses on the battlefield. Although Moctezuma was a valorous and able field general he was also a priest, much inclined to introspective brooding and meditation, much swayed by what he interpreted as omens. Once during a military campaign he abruptly gave orders to his counselor to return to Tenochtitlan and put to death the tutors of his many children and the duennas of his many wives. It apparently was dictated by a vision. At his coronation more than five thousand captives had been sacrificed and their flesh eaten by the assembled royal and noble guests. Then, wrote Fray Diego Durán, one of the ablest of the chroniclers, "everyone went to eat raw mushrooms. With this food they went out of their minds . . . worse . . . than if they had drunk a great quantity of wine. . . . With the strength of these mushrooms they saw visions and had revelations about the future. . . ." Such occasions were called Feasts of Revelations. When the hallucinogenic effects had worn off, the guests were expected to compare notes on what they had seen of the future while under the influence.

In view of what happened later it appears likely that many of Moctezuma's fears and preoccupations were concerned with an old deity named Quetzalcoatl, the Plumed Serpent. He was also known as the god of wisdom and enlightenment, associated with the wind and the morning and evening stars. He may have been either a god or a priest-ruler who assumed the god's name. Quetzalcoatl had been worshipped in the ancient city of Teotihuacan and he apparently reached a climax of influence during the Toltec era.

Quetzalcoatl and his followers abhorred human sacrifice. Quail, yes. Snakes, yes. Small animals, yes — but humans, no. This attitude put them in conflict with the principal deity of the Toltecs, Tezcatlipoca, or Smoking Mirror. Followers of Tezcatlipoca indulged in human sacrifice — not to the extent that the Aztecs later did in honor of Huitzilopochtli — but still enough to cause friction with the peaceful, life-loving followers of Quetzalcoatl.

This respect for life was not the only benign aspect of Quetzalcoatl. Tula was his city, and there he invented and encouraged useful arts and crafts, cultivated valuable crops and stimulated the love of the beautiful. Nahuatl myths reconstructed by post-Conquest scholars described an idealized Tula at the time of Quetzalcoatl. Everyone was happy and all things were abundant. Pumpkins were so large that they could hardly be lifted, and ears of maize were not only huge but so plentiful that they lay scattered about the ground and were used as fuel for fires. Cotton grew in many colors, ready to be spun and woven into handsome garments. Cacao trees flourished and the chocolate was the finest. Gold and jewels were so plentiful that no one paid them any heed. Birds with brilliant feathers — blue, green, yellow, red — sang beautifully. The people had no needs and knew no sadness. But it was the sort of paradise that could not last. Evil spirits tried to persuade Quetzalcoatl to kill men in sacrifice. Failing, they tricked him.

The "demons" first persuaded Quetzalcoatl to look in a mirror and become aware of his own ugliness. Then they dressed him in fine costumes and jewelry and adorned his face with a false beard of red and blue macaw feathers. Aware of his improved appearance, he abandoned his life as an ascetic and went out in the world in his new finery, a victim of vanity.

Next the demons gave him *pulque,* the wine of the country, to drink. He became drunk, sent for his sister, the priestess Quetzalpetlatl, and committed incest with her. When he sobered up Quetzalcoatl was overcome with remorse. He destroyed his fine houses, turned the precious cacao trees into cactus and mesquite, hid his wealth and fled. He traveled toward the eastern seacoast, followed by all the beautiful, colorful birds. On the seacoast he boarded a raft of snakes and sailed off toward the east, promising to return another day.

It may have been nothing more than a symbolic tale of the collapse of the Toltec empire, but it left its mark on succeeding cultures, including the Aztec. The Aztecs added Quetzalcoatl to their pantheon. But they corrupted the old life-loving god; they portrayed him with garlands of skulls and baskets of human hearts. Still, the tradition of Quetzalcoatl's benignity and his promised return persisted and must have been a particular problem for the suspicious, troubled mind of Moctezuma.

A neighboring ruler, Nezahualpilli, the sorcerer king of Texcoco, fed these suspicions. Nezahualpilli had a grievance. Moctezuma's father had given Nezahualpilli one of his daughters as a bride. The bride, little more than a child, turned out to be a nymphomaniac who publicly shamed the Texcocan king with an astonishing number of lovers. The bride, her lovers and her servants — thousands in all — were publicly burned, but the desire for vengeance must have lingered in Nezahualpilli's mind, for he called on Moctezuma and told him of omens of coming disaster: "In a very few years our cities will be ravaged and destroyed. We and our children will be killed and our vassals belittled. . . . You will see that whenever you wage war you will be defeated . . . before many days you will see signs in the sky."

The signs in the sky were not long in coming. One night a young attendant in the temple of Huitzilopochtli arose to relieve himself and was startled to see a huge three-tailed comet in the eastern sky. Moctezuma was informed and mounted a personal sky watch. The comet blazed from midnight to dawn for forty nights. Moctezuma called his astrologers and demanded an explanation. When none was forthcoming he ordered the astrologers killed. He fed his priests hallucinogenic mushrooms in the hope of visions of the future. The priests had none. He summoned Nezahualpilli for another conference. The latter made even direr predictions of catastrophes to come, adding: "Not a thing will be left standing. Death will dominate the land. All your dominions will be lost . . . it will all happen in your time." Some versions of the conversation credit Nezahualpilli with the prediction that the comet signaled the return of Quetzalcoatl.

Whatever it was, it deeply troubled Moctezuma, and omens and portents seemed to multiply. A boulder was being brought from Chalco to serve as a sacrificial stone in the great temple. The rock spoke to the workmen who were trying to move it: "O wretched people, O unfortunate ones, why do you persist? I will never arrive in Mexico. Go tell Moctezuma that it is too late . . . he no longer will need me; a terrible event, brought on by my fate, is about to happen! Since it comes from a divine will he cannot fight it. . . . Let him know that his reign has ended." Then the boulder broke through a bridge, fell in the water and could not be found again.

Fishermen found a strange, dark, cranelike bird in their net. In its head was set a mirror. It was taken to Moctezuma as a curiosity. Gazing in the mirror Moctezuma thought he saw a starlit sky and the images of strange warriors, marching and fighting.

Flocks of birds blotted out the sun. When shot down with arrows, their bodies were found to contain only dust. Men with two heads were seen walking in the street. One of Moctezuma's sisters, recently dead, arose from the grave to warn her brother of impending doom. The great temple caught fire and burned. Water thrown on it only made the flames leap higher. A lone woman walked the streets at night crying, "O my sons, we are lost. . . . O my sons, where can I hide you?"

An old woman told of her dream. She had seen "a mighty river enter the doors of your palace, smashing the walls . . . the temple too was demolished . . . great chieftains and lords filled with fright, abandoning the city and fleeing toward the hills."

An old man refused to reveal his dreams and was thrown in prison. He told his jailer: "Let Moctezuma know, in one word, what is to become of him. Those who are to avenge the injuries and toils with which he has afflicted us are already on their way. I say no more." So ran the ominous legends.

The signs and portents were said to have stretched over a ten-year period prior to the arrival of Cortés and his men in Mexico. One way or another Moctezuma seemed to have become convinced that Quetzalcoatl was in fact returning, just as had been predicted.

Even before the Spaniards reached Mexico there had been rumors in Tenochtitlan of strange, fair-skinned men in the islands to the east. Moctezuma maintained an excellent intelligence network. Traveling merchants and tribute collectors were required to observe carefully anything and everything untoward that happened in the vast Aztec realm. They were often accompanied by artists skilled at depicting people and events in drawings which would then be carried to Tenochtitlan by fleet runners. Word of Córdoba's incursion in Yucatán had reached Moctezuma in this way. So had information about Grijalva's visit to Tabasco. When Grijalva and his men reached San Juan de Ulúa, Moctezuma had envoys on hand to greet them with gifts of gold, jewels, cotton mantles and food and to learn what they could of the strangers' nature and intentions. Grijalva, lacking imagination, curiosity and any way of communicating with these strange people, merely thanked them, sent the gifts back to Cuba and proceeded on his way up the coast.

Then came Cortés.

SIX

Veracruz (I)

WHEN CORTÉS AND his men dropped anchor at San Juan de Ulúa they were almost immediately visited by canoeloads of natives. Having observed which was the most impressive of the eleven vessels, the Indians came aboard Cortés's flagship. Their demeanor was respectful. They scraped the sea slime from the bow of the ship and touched it to their lips.

For this first occasion the only interchange was ceremonial and symbolic. Cortés attempted to tell them that he had only come to trade with them and meant no harm — but his listeners could not have understood. Cortés presented them with the usual baubles and also gave them Spanish food and wine to drink. They were particularly taken with the wine, and by sign language indicated that they would like to take some of it to their chief.

The next day the Spaniards disembarked and exercised their horses in the sand dunes that stretch along the coast. It was not a good campsite. Wind shifted the dunes and made the footing difficult for both men and horses. When the wind swung to the north it was cold. When the wind dropped there were clouds of biting, stinging insects. There were many stagnant swamps and a smell of decayed marine life. Indians volunteered to help the Spaniards set up camp, building shelters of tree limbs and thatch. They also brought welcome provisions of roast turkeys, smoked fish, tortillas, clay pots full of beans and a wide variety of tropical fruits. Despite the natives' friendliness Cortés set up his artillery to protect the campsite from possible attack. Meanwhile it had been discovered that Marina could understand the language of these people and, with Aguilar's help, translate for Cortés.

On Easter Sunday a vast procession of Indians — three to four thousand of them — approached the camp. All of the Indians were finely dressed in flowing garments, and none of them appeared to be armed. They were led by two great chiefs. One of them, Teuhtlili, was governor of the province of Cuetlaxtla, in which the Spaniards were camped. The other was Cuitlaltipoc, who had functioned as ambassador for Moctezuma in presenting gifts to the Grijalva expedition the previous year. The attitude of Teuhtlili, Cuitlaltipoc and their followers was both dignified and reverential. They bowed to the ground, touched their fingers to the sand and then touched their mouths. They swung clay censers with pungent copal incense around Cortés and the captains who stood near him. They brought vast supplies of food; much of it was sprinkled with human blood, pricked, with cactus spines, from the earlobes, tongues and lips of the food bearers as a mark of special piety. Other bearers brought food for the horses, more turkeys, tortillas and bunches of flowers. There were special gifts for Cortés, the leader, and the Indian envoys adorned him with them: a turquoise mosaic mask in the form of a serpent's head; a headdress made of green quetzal feathers; green stone earplugs in the form of coiled serpents; a necklace of green stones supporting a golden disk; a polished stone mirror to be worn in the small of the back; leg guards made of green stone beads and golden shells; sandals made of obsidian — impossible to walk in but highly decorative; a sleeveless jacket; a voluminous cape and a shield decorated with gold, seashells and quetzal feathers.

Cortés, seated on a chair like royalty, accepted the gifts with a mixture of dignity and pleasure; he must also have been puzzled by the ceremony. He retaliated with a ceremony that must have been equally puzzling to his visitors: he had Fray Bartolomé de Olmedo conduct a special Easter mass. The visiting Indians followed the proceedings with gravity and bafflement.

After the mass Marina translated, with Aguilar's help, a message from Cortés to his visitors. It was a somewhat more complex statement than the one Cortés had made a few days earlier, when he said he had come only to trade. The Spaniards were subjects of the most powerful king in the world, a monarch whom many kings and princes served as vassals. This monarch had long known of Moctezuma's greatness in his own land and had sent Cortés and his men across the sea to enter into friendly relations with Moctezuma, to present him with gifts and to convey friendly messages. This they must do in person, in the presence of Moctezuma.

All of this was bold and wildly imaginative. The king, a Hapsburg, who was both Charles V of the Holy Roman Empire and Carlos I of Spain, knew nothing of Moctezuma. It is unlikely that he knew anything of Cortés, and almost certainly he did not know where he was or what he was doing. Charles had come to the throne three years earlier, in 1516, upon the death of his grandfather, Ferdinand of Aragon, who had ruled alone in Spain during the twelve years since the death of Isabella of Castile. The throne was to have gone to their daughter, Joanna of Castile. But the heiress, known as *la loca,* or "the mad," was certifiably deranged, and so the throne had gone to her son, Charles of Hapsburg, to rule in her name. The young king's extravagances would almost bankrupt Spain, and his chronic need for gold would play a key role in the fortunes of Cortés and his men.

Teuhtlili, the native spokesman, was taken aback by Cortés's insistence upon an audience with Moctezuma. "You have hardly arrived here," he replied, "and already you want to speak to our emperor." Nevertheless, he summoned bearers with additional gifts for the Spaniards: gold trinkets, gem-stones, featherwork and finely embroidered cotton cloths. Cortés responded with gifts to be conveyed to the great Mocte-zuma: a carved and painted wooden armchair, a red cloth cap adorned with a gilt medallion of St. George slaying the dragon, various other articles of Spanish clothing and neck-laces of glass beads. He also commanded his little cavalry troop to stage a display of horsemanship. He had one of the cannon loaded with an extra charge of powder so that it would throw a stone ball far back into the dunes. While all of this was going on, Aztec artists who accompanied Teuhtlili were busily sketching Cortés, his men, the horses, ships and firearms, limning them skillfully on sheets of rough bark paper.

Teuhtlili promised to have messengers convey to Mocte-zuma Cortés's request for an audience. But before leaving he noted that one of the Spanish soldiers wore a battered gilt helmet which was, he said, very like the helmet worn by the Aztec god Huitzilopochtli. He asked if it could be taken to show Moctezuma. Cortés gladly agreed. And it occasioned one of his most prophetic remarks. The helmet could be taken to the Aztec ruler and examined provided that it be returned filled with gold. For, he said, "We Spaniards suffer a disease of the heart which can only be cured with gold." It was a state-ment which might have seemed innocuous at the time. The Aztecs had no particular regard for gold except for its decora-tive quality. They referred to it as "excrement of the gods." Jadeite was considered far more precious, and their unit of exchange was the simple cacao bean.

Teuhtlili had been sending daily couriers to Moctezuma with verbal dispatches and illustrative pictures drawn by the artists. He assured Cortés it would not take long to get an answer to his request for a personal visit. Teuhtlili departed with his retinue, leaving behind Cuitlaltipoc and several thousand Indians charged with building adequate shelters for the Spaniards and providing them with food. Uncomfortable as the location was, the Spaniards felt at ease for a little while.

Teuhtlili returned to Moctezuma's capital, Tenochtitlan, a difficult trip through jungles, mountains and deserts of several hundred miles. His monarch became greatly agitated and more than ever convinced that the white invader either was or represented the long-missing god Quetzalcoatl. He consulted advisers, astrologers, sorcerers and allied princes. One of the latter, Cuitlahuac of Ixtapalapa, stoutly advised him to allow the Spaniards to go no farther; if permitted to do so, he insisted, the foreigners would destroy both the empire and Moctezuma himself. Moctezuma, while generally in agreement, still did not want to offend these strange creatures who just might be divine. He dispatched Teuhtlili to return to the coast with still more lavish gifts for Cortés and his men.

The newest assortment of gifts from Moctezuma was the richest the Spaniards had yet seen. There was the helmet filled with flakes and nuggets of gold, as Cortés had requested, golden figurines of animals, birds and fishes, necklaces and medallions and, wonder of wonders, two huge disks, each the size of a cartwheel and several inches thick. One was of silver, with engraving indicating that it represented the moon, the other of solid gold, representing the sun.

It was a splendid haul — but the words brought from Moctezuma were discouraging to Cortés. Moctezuma said he had followed with interest the reports of white men coming to his country over the past several years, and he was pleased to learn that they came as envoys of a great monarch across the seas, whose friendship he was happy to acknowledge. The Spaniards must feel at home in his country; he would be happy to provide them with anything they might need. But a personal interview was out of the question. He himself was too ill to leave his capital; and the country was so difficult for travel that Cortés and his men must not even consider trying to make the journey inland.

Once more Cortés indulged in the sort of flexible untruth at which he was becoming so adept. He said that he had traveled thousands of miles across the ocean in order to deliver in person the king's messages. He would not be deterred by a difficult journey of a few hundred miles. Indeed, he would not dare to return to Spain without having performed the task with which he had been commissioned. Teuhtlili unhappily agreed to convey Cortés's response to Tenochtitlan, and once more departed.

Teuhtlili made one more appearance. He arrived with a gift of green stones for Cortés and a final message from Moctezuma. The Spaniards must leave this country. There would be no more discussion of the matter. The envoy's manner was firmer than it had been. The Indians, through sharp observation, had discovered that the Spaniards were more mortal than divine. They had normal appetites for food and women. Their excrement was not gold but like their own.

After a few more days the several thousand Indians who were attending the Spaniards' camp disappeared. One day they were there, bringing food for the men, fodder for the horses. On the next they were gone and the Spaniards were alone, reduced to catching fish and crabs for their food. This could be supplemented only by raiding Indian villages — all of which appeared to be deserted — for supplies of maize. The weather was growing warmer — unbearably hot in the middle of the day — and the insects fiercer. The campsite, of doubtful utility even with the services of the Indians, was impossible without their help. Cortés dispatched one of the ships, commanded by Francisco de Montejo and piloted by the veteran Alaminos, to explore the coast to the northwest, looking for a safe anchorage, a healthier climate, ample supplies of sweet water and access to arable land. The rainy season would soon be upon them. Great white cumulus clouds shrouded the peaks of the mountains to the west, obscuring the great snow-clad summit of Citlaltepetl (Orizaba Peak), which usually could be seen from the coast, catching the rays of the morning sun while all the rest of the land was dark.

But a suitable campsite was not the only problem. From the

beginning of the expedition the Spaniards had been divided into anti- and pro-Velázquez forces. Movement, activity and Cortés's skill as a leader had until now minimized these differences. Now, however, they broke into the open, and many of the Spaniards insisted that it was time to take the treasure they had received from Moctezuma's envoys and return to Cuba.

Cortés and the men close to him, on the other hand, insisted that they must make a settlement where they would be secure from hostile acts by the Indians and from which they could explore and learn more of this country and its great riches. This appeared to have been what Cortés had in mind all the time. Since he was already in violation of Velázquez's instructions on many counts he knew that any help from Cuba was not likely — and help from the mother country, Spain, was even less so. He was, quite literally, cut off from any support or communication. He needed a base of operations. He began discussing his — and their — plight with some of his most trusted followers: the Alvarado brothers, Puertocarrero, Cristóbal de Olid, Alonso Dávila, Francisco de Lugo and Juan de Escalante. The idea was planted that they should establish a *villa,* or town, which would be responsible directly to the Spanish crown rather than to the colonial authority of Velázquez. Cortés's trusted friends began propagating the idea, persuading others who were presumably friendly to Cortés, or at least not openly committed to Velázquez, that this was the only logical course. Finally they approached Cortés, suggesting as their own idea the course that Cortés had already chosen to follow. They insisted that he must name officers for the town-to-be. This he did, giving some appointments to Velázquez men but the more important ones to his own allies. Thus a *cabildo,* or town council, was formed without, as yet, a town for it to govern, although it had already been given a name: Villa Rica de Veracruz, the rich town of the true cross.

There then followed a feat of legal humbug that Cortés performed masterfully. He submitted to the *cabildo* the instructions he had received from Diego Velázquez, including the prohibition against settlement and the order that after the expedition had done as much trading as possible it should return to Cuba. At the same time, he submitted his resignation as leader of the expedition. The *cabildo* then declared the town a direct dependency of the Spanish throne and appointed Cortés as captain-general and chief justice — an officer of the crown rather than a subordinate of Diego Velázquez, lieutenant governor of Cuba. The land was claimed for Spain; Grijalva's earlier claim was regarded as invalid since he had claimed it in the name of Velázquez. It was all somewhat farcical, but it was the legal or quasilegal foundation on which the coming conquest of Mexico was to be built.

Now all that was needed was a site to build a town for the *cabildo* and the newly appointed captain-general and chief justice to govern. Francisco de Montejo returned from his exploratory trip along the coast and reported that the only suitable site was about twenty-five miles to the northwest. There a huge rock protected a suitable anchorage from the treacherous north wind. There were freshwater streams, ample wood and stone for buildings, and fertile land. There was, nearby, an Indian town, Quiahuiztlan, which could supply food and a labor force.

Despite some grumbling from the pro-Velázquez faction, camp gear, supplies and artillery were loaded on the ships and dispatched to the new site. Cortés and his men marched overland. At the site Cortés, as ill-disposed toward hard physical labor as any Spaniard, set an example for all his men, digging, hewing logs, carrying stones and earth. Within a few weeks something resembling a town had arisen: a church, an arsenal, a storehouse, a stockade, a slaughterhouse, a meetinghouse for the *cabildo,* a plaza, a prison and a gallows.

Cortés had kept some of the dissidents out of the way by sending them off on a foraging expedition led by Pedro de Alvarado. Others who were unmanageable any other way were imprisoned on the ships on the orders of the captain-general and chief justice. Others, such as Velázquez's kinsman Juan Velázquez de León, were won over to Cortés's side by flattery, bribery or both.

The first fortnight of July was to be a watershed period in the history of the minuscule Spanish outpost. On July 1 a ship arrived. It was a caravel which Cortés had bought in Santiago but had left behind for repairs when the expedition set out from Cuba. It was commanded by Francisco de Salcedo. Salcedo brought with him welcome food supplies, one horse, ten soldiers and a captain, Luis Marín. He also brought news of utmost importance. Diego Velázquez had been successful in his negotiations: he had been named *adelantado*, or governor, of Cuba — until then his title had been that of lieutenant governor, a subordinate of the colonial government in Española. Moreover, he had received powers to both trade and establish a settlement in Yucatán, a name that was indiscriminately applied to all the mainland territory that had been found by Córdoba, Grijalva and Cortés.

The news set off a flurry of purposeful activity in Villa Rica de Veracruz. Velázquez partisans took heart and became more belligerent, demanding an immediate return to Cuba. The most vocal of the malcontents were seized. Two of them were hanged. One of them was Juan Escudero, the man who had once arrested Cortés on Velázquez's orders in Cuba. Another had his feet cut off. Others were flogged and put in chains. Later, in a dispatch to the king, Cortés explained, somewhat blandly, "On hearing the confessions of these miscreants, I punished them according to the law and as, in the circumstances, I judged would do Your Majesty greatest service."

Cortés also prepared the best of his ships, the flagship, to sail directly to Spain with the king's share of the treasure thus far accumulated. In charge of the voyage he placed Francisco de Montejo, nominally a Velázquez man, and Alonso Hernández de Puertocarrero, Cortés's friend, confidant and "owner" of the handsome and talented Marina, whom Cortés was beginning to appreciate more and more.

Cortés compiled a letter to the king and his mother, the queen (Joanna, although unable to rule, was still recognized as queen, although her son Charles was performing all the royal functions). The letter was lost, but its contents were undoubtedly reflected in a letter written in the name of the *cabildo* and signed by all the Cortés partisans. It was extravagant in its praise for Cortés's enterprise and unsparing in its condemnation of Velázquez, who, it was suggested, would like to reap all the benefit of the Cortés expedition for himself. It begged that the crown not "give or grant concessions to Diego Velázquez . . . or governorship . . . or judicial powers; and if any shall have been given him, that they be revoked. . . . Were the aforementioned Diego Velázquez granted some office, far from benefiting Your Majesties' service, we foresee that we, the vassals of Your Royal Highnesses, who have begun to settle and live in this land, would be most ill used by him, for we believe that what we have done now in Your Majesties' service, namely to send you such gold and silver and jewels as we have been able to acquire in this land, would not have been his intention. . . ." The document also stated that the pro-Velázquez conspirators who were being punished had wanted to seize all the valuables thus far obtained and send the treasure to Velázquez.

The letter added that citizens of the town begged that Cortés be appointed "Captain and Chief Justice of Your Majesties and to give him . . . royal seal and license that he may continue justice and good government amongst us. . . ."

This was all partisan argument. Cortés arranged a more convincing argument in the treasure itself. By rights he was entitled to one-fifth of the *montón*, or booty, after the royal fifth had been subtracted. This would go toward reimbursement for his expenditures in securing ships and supplies for the expedition. He not only foreswore his share but persuaded his comrades to give up theirs also, so that the entire treasure would go to Spain. There was much more wealth to be obtained in this vast country, he suggested, and in time they would all be rich men. Besides, much of the treasure, the fine fabrics, the beautiful featherwork, some of the intricately wrought ornaments, could not easily be divided. Although there were some complaints from Velázquez supporters, Cortés's suggestion was approved.

Dominating the treasure hoard were the two great wheels, one of gold and one of silver. There were also a gold necklace set with both green and red stones and pearls and hung with gold bells; a gold bracelet; a wand or scepter girdled with gold and pearls; a wooden headdress decorated with gems and golden bells and surmounted by a brilliant green bird with eyes, beak and feet of gold; tridents decorated with feathers and tipped with gold and pearls; two dozen golden shields decorated with feathers and pearls and other shields of silver; fish, ducks, birds and seashells, all carefully modeled and cast in gold. And although Cortés had announced that he was giving up his share, there was also a sizable parcel of gold to be delivered to his father.

Cortés ordered the treasure ship not to stop in Cuba, but the orders were disobeyed because Montejo wanted to look at some property he owned. Eventually word of the ship reached Velázquez in the eastern end of the island, and he ordered a caravel out to overtake it and capture the treasure. By this time, however, Montejo and Puertocarrero, guided by the pilot Alaminos, had sailed out into the Bahama channel and were on their way to Spain.

Cortés was determined to march to the interior, but he had one more dramatic step to take before doing so. He must cut off all possibility of retreat for those who, through either cowardice or disloyalty to him, were not in favor of the venture. There were still ten ships that offered the possibility of retreat. They must be destroyed.

He persuaded some of the sailing masters to bore holes in the ships' hulls so that they leaked faster than the holes could be caulked. He encouraged some of the pilots to spread the word that all of the ships were being destroyed by sea worms and dry rot from having been so long in port. There was much discussion of what should be done. As Bernal Díaz del Castillo put it, "Cortés had already decided to burn the ships but wanted it to appear that the idea came from us. . . ." Actually they were not burned, but they were damaged beyond use.

Cortés ordered that the ships be stripped of whatever was salvageable — sails, rigging, guns, provisions, anchors, hardware. Five of the best ships were allowed to drift ashore. Four others were then sunk in the harbor. Some of the dissidents now approached Cortés and accused him of condemning them to death in this alien land. Cortés chided them for being unwilling to risk a little danger for the sake of great riches, but added that he had saved one ship in order to take the fainthearted back to Cuba. With this ruse he was able to make up a list of those not to be trusted in the future. As soon as he had satisfied himself on this, he ordered the last ship scuttled.

During all this maneuvering, Cortés had been gathering intelligence about the native population of the country he intended to conquer. While still at San Juan de Ulúa, waiting out the exchange of messages with Moctezuma, Cortés had been visited by Ixtlilxochitl, one of the sons of the late Nezahualpilli of Texcoco, the sorcerer king of Texcoco, Moctezuma's neighbor who had warned the susceptible Aztec ruler of dire events to come. After Nezahualpilli's death Moctezuma had been influential in putting another son, Cacama, on the Texcocan throne. Ixtlilxochitl had become, as a result, a plotter against Moctezuma. Learning of Cortés's presence, he came to the Spanish camp, represented himself as an enemy of Aztec tyranny and proposed an alliance. This may have been Cortés's first hint of the disunion and unrest that existed in the Aztec empire.

And there was soon more proof of it. A little after the Spaniards had been deserted by the Aztec servants and food providers, Cortés was approached by a small group of Indians who came from the province of Totonacapan, a region a little to the northwest of the campsite at San Juan de Ulúa. Their language differed from the Nahuatl spoken by the Aztecs, but the ingenious Marina discovered that a few of them spoke Nahuatl and was thus able to speak with them. Totonacapan, they said, had once been an independent kingdom, rich in history and culture. But it had in recent years been forced to pay tribute to the Aztecs and resented having to do so. They invited Cortés to visit their chief, Chicomacatl, in the capital of the province, Cempoala. On the overland march from the dunes of San Juan de Ulúa to the harbor and townsite found by Montejo, Cortés and his army did so.

The march took them across several rivers, through rich farmlands separated by jungles full of orchids and brilliant-colored birds. They passed through several deserted villages whose pyramidal temples displayed grisly evidence of recent human sacrifice. When the horsemen riding ahead of the marching column first sighted Cempoala, one of them dashed back to report that the buildings had walls of silver; actually they were covered with white stucco.

The Spaniards were welcomed into the city with bouquets of flowers and clouds of incense. Chicomacatl, the chief, was so grossly fat that he could walk only when supported on either side by companions — and the Spaniards thereafter referred to him as the Fat Chief. Comfortable quarters with freshly white-washed walls and abundant supplies of food were provided for the Spaniards. Soon the Fat Chief was explaining his woes to Cortés — the cruelty of the Aztecs, their constant demand for tribute, their seizure of Totonacs to be sacrificed on the altars of Tenochtitlan. Cortés listened with interest and sympathy. Through his complicated linguistic chain — Spanish to Mayan to Nahuatl to Totonac — he gave his set speech: that the Spaniards were vassals of a great and powerful king, Don Carlos, who had sent them to seize wicked persons, to punish them for their evil deeds and to put an end to human sacrifice. He also explained that he was marching to Quiahuiztlan, not far away, where his ships awaited him. Once settled there they could talk some more about what should be done about the Aztecs.

The Fat Chief provided several hundred porters to assist the Spaniards on their march, to carry heavy freight and supplies. With their help the Spaniards quickly marched on to Quiahuiztlan, situated on a steep mountainside overlooking the harbor in which the ships were moored.

Although most of the townspeople of Quiahuiztlan had fled in fright, a few of the principal men had stayed behind and they welcomed the Spaniards in friendly fashion. The Spaniards had barely had time to catch their breath when the obese chieftain of Cempoala arrived also, borne in a litter by a party of sweating slaves. Then a third group arrived: five haughty men, elegantly gowned and adorned with jewels, carrying bouquets in their hands and attended by servants with fly whisks to discourage the mosquitoes. They were escorted to one of the most elaborate dwellings in the town and presented with a feast. This they seemed to take for granted; no word was spoken and the newcomers stared through the five hundred or so Spaniards without appearing to notice them.

These, Cortés learned, were the Aztec tribute collectors, come to collect the regular exaction of twenty persons who would be transported to Tenochtitlan as sacrificial victims. While the tribute collectors busied themselves with their feast, Cortés went on parleying with the Fat Chief and the principal men of Quiahuiztlan. Once more Cortés explained his mission — to end injustice, avenge wrongs and bring peace and prosperity to this distant corner of Spain's world. This brought another flood of complaints about the cruelty of the Aztecs; some fifty towns in Totonac territory were similarly afflicted. Cortés assured them that he had come to put an end to such tyranny; he would ally himself with the long-suffering Totonac people to avenge these wrongs. And, in a remarkable demonstration of intuition, he suggested that the natives should celebrate their new alliance by not only refusing to pay tribute but also by seizing and imprisoning the haughty tribute collectors.

The suggestion at first brought only exclamations of alarm

and dismay. But Cortés assured them that he and his army stood ready to reinforce them. Finally the natives agreed. The tribute collectors were interrupted at their feast, were tied hand and foot to long poles and imprisoned. The Totonacs, having turned from reluctance to enthusiastic activism, proposed to kill the hated Aztecs, but Cortés restrained them. Meanwhile, couriers carried word of the open rebellion to all other Totonac towns.

Cortés and his men went on to their new townsite near the anchorage, a few miles away, and began work on the town-to-be. After darkness fell Cortés sent a raiding party back to the house in Quiahuiztlan where the tribute collectors were imprisoned. The guards were asleep. Two of the prisoners were taken back to Spanish headquarters. Cortés displayed his skill at dissimulation. He professed not to know who the prisoners were or why they had been brought before him. He knew, of course, who they were, and suspected that their earlier disdain for the Spaniards in Quiahuiztlan had been on direct orders of Moctezuma, who seemed to have switched from cordiality to hostility.

The prisoners, now unbound and being served a meal of Spanish food, insisted that Cortés did indeed know who they were and that the Totonacs would not have dared to defy the will of Moctezuma had Cortés not incited them to do so. This, Cortés insisted, was wrong. To prove it he intended to set them free. A Spanish boat would take them down the coast, out of Totonac territory, and there they would be released. They must hurry to Tenochtitlan and convey assurances to Moctezuma that he, Cortés, was his devoted friend who had nothing but admiration and goodwill for the Aztec ruler.

In the morning Cortés upbraided the Totonacs for allowing two of the prisoners to escape, and insisted on taking custody of the remaining three. They were put in chains and taken aboard one of the ships. Meanwhile Cortés met with the *caciques* of Cempoala and Quiahuiztlan and other nearby Totonac settlements. The *caciques* were well aware of the gravity of the crisis, and two courses were debated. One was to throw themselves on the doubtful mercy of the Aztecs, pray for forgiveness and hope for the best. The other was to rally all Totonac communities in open rebellion and, with the help of the Spaniards, throw off the oppressive rule of the Aztecs and regain their independence. Cortés listened carefully. The Totonacs claimed they could field an army of one hundred thousand men. It was an impressive figure — which turned out to be exaggerated. Cortés urged on them, as a condition of his aid, that they not only throw off the Aztec domination but that they also acknowledge themselves to be vassals of the king of Spain. This they agreed to do and the fact was legally recorded by Diego de Godoy, the notary.

When word of the Totonac rebellion reached Moctezuma, he had ordered an army prepared to march to the coast to punish the insurgents, restore order and destroy the army of invaders. But before the force was ready to leave the two liberated prisoners arrived with their tale of the Spanish leader's kindness, generosity and solicitude. Moctezuma, as he was so often to do in the future to his own misfortune, changed his mind. Instead of a punitive expedition he sent off a small embassy of trusted councillors, including two of his nephews, to meet Cortés on the coast. They carried, as usual, an assortment of gold and other gifts.

Cortés received the embassy with cordiality and gave reassurances of his esteem and admiration for Moctezuma. As proof of his goodwill he turned over to the envoys the three tribute collectors whom he still held prisoner. Moctezuma's men accepted this gesture with good grace but insisted that the Totonacs must be punished for their insurgency and insubordination. Cortés politely told them that this was out of the question: that the Totonac region was no longer subject to Aztec domination, since the various chieftains had now declared themselves vassals of the king of Spain. It was a complicated matter and one which Cortés himself hoped to explain in person to Moctezuma, just as soon as a meeting could be arranged. The emissaries departed for Tenochtitlan with this perplexing information. Moctezuma pondered the matter. The mixture of deference and defiance was baffling. His priests and soothsayers could not interpret it. Skilled sorcerers whom he had ordered to haunt the Spaniards' camp reported that the strangers seemed magically immune to their most powerful spells. Were they men or gods? One chronicler recorded Moctezuma's reaction to the sorcerers' failure: "You have done everything in your power. . . . Perhaps when the strangers arrive here your enchantments and power over dreams will be more effective. Let them enter the city, since it is here that we will find a way to destroy them totally."

Cortés meanwhile continued to demonstrate his capacity to make the most of his opportunities. When the Fat Chief of Cempoala asked his military support in attacking a Mexican garrison at the nearby town of Cincapacinga he agreed. The supposed garrison town was attacked, but it was quite apparent that it was nothing more than a peaceful Indian community; the Fat Chief wanted only an opportunity to loot, which his warriors proceeded to do.

Cortés realized that he, the manipulator, had been cleverly manipulated by his newfound Indian allies. He ordered a retreat. Once back in Cempoala, the Fat Chief, repentant, presented to Cortés eight young women to give to his men. The chief's own niece — a fat and homely girl with a considerable dowry of jewels and finery — was for Cortés himself. Cortés took the occasion to function as a missionary. The Spaniards could not accept the girls until the girls had been baptized as Christians. Furthermore, the town's heathen idols must be destroyed.

The Fat Chief objected. To do so would bring down the wrath of the gods on their heads; any such move would be resisted to the death. Cortés responded by having the Fat Chief and several others seized. Marina reasoned with them. If the Cempoalans refused, they would lose the support and protection of the *teules,* or gods: that is, the Spaniards. If there was any resistance, the chiefs would be put to death on the spot. Meanwhile a party of fifty Spaniards mounted the steps of the pyramid and entered the temple. Idols were thrown down the steep sides. The Cempoalans watched aghast. Then the Spaniards cleaned the bloodstained temple, whitewashed the walls and erected a wooden cross. Fray Bartolomé de Olmedo delivered a sermon and instructions for care and veneration of the Christian shrine, how it should be regularly supplied with fresh flowers and wax candles. Some of the Totonac priests had had their blood-clotted hair washed and shorn. Then, dressed in new white robes, they carried lighted candles to the shrine at the top of the pyramid and prostrated themselves before the cross and an image of the Virgin Mary. The Cempoalans, already tied to the Spaniards by armed diplomacy, were now linked with them through the mystical — and probably ill-understood — bonds of religion.

In the process of allying himself with the Totonacs Cortés had accumulated considerable understanding of the country, its peoples and his own tactical situation. He understood that he and his men were all regarded as gods, or, if not gods, certainly supernatural beings with mysterious powers. He also had begun to understand the legend of Quetzalcoatl and the reason why his own person had come to be identified with that of the old god. It was a matter of potential significance. On a more pragmatic basis he sensed the deep schisms that ran through the Aztecs' empire, and that while he and his men might be feared, even hated, they would be no more feared and hated than were the Aztecs by the various subject peoples. He by now had committed himself to a march up to Tenochtitlan. His Totonac allies promised him they would provide him with porters to carry the expedition's supplies and equipment and also the best of their warriors to fight under his command. Furthermore, they said, he could expect to enlist even more native allies in such inland provinces as Huejotzingo and Tlaxcala. He might conquer an Indian empire with Indian troops. But, best of all, Moctezuma's vacillation led him to think that perhaps, with God's help, he might conquer an empire without firing a shot or shedding a drop of Spanish blood.

He garrisoned his new town of Villa Rica de Veracruz with one hundred men under the command of a trusted lieutenant, Juan de Escalante. And on August 16, 1519, with his little army of Spanish adventurers, Indian warriors and porters, he left Cempoala on his march to Tenochtitlan and, he hoped, gold and glory.

SEVEN

Tlaxcala

ON THEIR MARCH toward the unknown heart of Mexico, Cortés reminded his men of their precarious situation. They had no way to retreat. There was no hope of help from Cuba. They might someday be helped by the mother country. But at the moment the mother country did not know where they were or what they were doing for the glory of God and Spain. Their only hope lay in divine favor and their own stout courage. However, since their mission was undertaken in God's name and for the propagation of the true faith, divine providence would not fail them, nor would their own brave spirits falter. The men cheered. And with the ensign, Corrall, carrying the banner with the cross and the royal arms, the troop set off: four hundred Spanish foot soldiers, fifteen horsemen, a column of more than one thousand Totonac warriors commanded by three *caciques,* and a train of two hundred Totonac porters carrying the six pieces of artillery and other heavy equipment and supplies. Cortés rode at the head of the column.

For the first few days they advanced through the lush disorder of the tropics, following the course of the Chachalacas

River. The air was loud with the cries of birds and heavy with the fragrance of hot country fruits and flowers. Then, as the jungle thinned, they began to climb into the subtropical zone, where the flowers were still abundant and bright but where the trees — live oaks, cedars, pines and sweet-smelling gum trees — grew taller and more stately. Up and up they climbed, with the snow peak of Citlaltepetl on their left and a strange bleak mountain, topped with a black coffin-shaped rock, on their right. There were fast-running streams rushing toward the gulf they had left behind them, rocky ledges and narrow defiles where it was difficult for the horses to keep their footing. They crested the first mountain range and emerged on a broad and barren tableland where the remnants of ancient lakes had rimed the landscape with gypsum and salt. There were hailstones and bitter cold; most of the Totonac and Cuban servants suffered from the weather, for which they were ill-clad, and some of them died. Ahead lay more mountains, range after range, their slopes thickly forested, between them occasional green valleys and more deserts, usually peppered with that awesome American curiosity, the cactus.

In several villages they were received in friendly enough fashion and given simple food. Cortés would have the *requerimiento* read to the uncomprehending audiences and crosses would be erected.

After eight days of marching the column reached Xocotla, a large town with handsome white buildings and thirteen temples. The chief, who commanded twenty thousand vassals, thirty wives, one hundred concubines and two thousand retainers, was an enormously fat man, Olintetl, fatter even than the Fat Chief of Cempoala. The Spaniards called him the Shaker, for the slightest movement would cause his body to quiver like jelly. When Cortés asked him if he was a vassal of Moctezuma, Olintetl replied, "Who is not?" He was impervious to the *requerimiento* and Cortés's talk of a Christian deity and Spain's vast dominions and greatness. Olintetl replied that Moctezuma, too, was great; that he commanded the loyalty of thirty great lords, each of whom could field an army of one hundred thousand men — a force of which staggered the imagination of the four-hundred-odd Spaniards who might have to oppose it. They were also awed by the sight of a criblike rack outside one of the temples, on which were arrayed human skulls by the thousands — one chronicler said at least one hundred thousand — victims of past sacrifices.

Olintetl was neither overly friendly nor generous with provisions. And he seemed unimpressed with Cortés's arguments that human sacrifice must be halted and the idols in the temples destroyed. Cortés later claimed in a dispatch to the king that he had destroyed the idols; actually the Mercedarian friar, Bartolomé de Olmedo, dissuaded him from attempting it as both a useless and a dangerous gesture.

Olintetl said that if the Spaniards were going to Tenochtitlan their best route lay through Cholula, a large religious center not far from the Aztec capital. Here they would find friendly people. The Totonac chiefs, however, counseled against it. The people of Cholula were treacherous, they said. The safest way was the one they had urged from the outset, by way of Tlaxcala, an independent province whose citizens hated the Aztecs as much as they, the Totonacs, did.

Olintetl had known of the Spaniards' coming and probably had been instructed by Moctezuma's couriers to receive the strangers. Still, his air had been one of cool arrogance, anything but the cordiality that Cortés had hoped for as a result of his intricate dealings with the Aztec emissaries in Totonacapan. And this may have figured in his decision to reject the advice about Cholula and go through Tlaxcala instead. Although his advice had been ignored, Olintetl provided a company of warriors to accompany the Spaniards — and perhaps spy on them.

In heavily wooded mountains beyond Xocotla the Spaniards found threads strung between pine trees and across the trail. From them hung bits of paper in strange shapes and bright colors. The Indian guides explained that these were charms to impede strangers from approaching Tlaxcalan territory. Cortés sent ahead four Indians as his agents to ask for safe and peaceful passage through Tlaxcala. Meanwhile the Spaniards moved cautiously ahead. They emerged from the mountains into a long valley, six miles wide. Crossing the nearest end of it was a huge stone wall, half again as tall as a man and nearly twenty feet thick, with battlements at regular intervals. In the center there was an overlapping entrance through which only a few men could squeeze at a time, easily controlled if some-one was defending the wall. But no one was, and the long column filed through into the province of Tlaxcala. Although the wall was undefended the Spaniards were, in a matter of minutes, involved in a deadly skirmish with a small but cou-rageous band of warriors.

The Tlaxcalans, whose territory this was, had heard of the Spaniards and had known of their alliance with the Totonacs, who were, at least in theory, their friends. But they also knew of Cortés's dealings with Moctezuma's agents and emissaries and suspected that all of this was somehow a trick of the hated Aztecs, with whom the Tlaxcalans had been at war for almost a century.

Tlaxcala and the Aztec empire had common roots. Both were descended from the savage northern tribes that had come down into central Mexico in the twelfth century. Both spoke Nahuatl and they worshipped common gods, under different names, paying homage in human sacrifice and ceremonial cannibalism. But the Tlaxcalans had shut themselves up in a small — fifteen by twenty-five miles — valley to the northwest of Tenochtitlan and had fiercely maintained their indepen-dence despite the Aztec domination of all the surrounding territory. Periodically there were devastating wars with the empire-building Aztecs, but the Tlaxcalans managed to sur-vive. Their country was fertile; the name meant "land of bread." There was water from the Atoyac River and there were broad fields of maize and maguey, the all-purpose plant that provided food, drink, building material and fiber. Their land was at too high an elevation to grow cotton or cacao, and they had no sources of salt; the Aztecs prevented them from getting any of these staples by trade with other regions. Their society was oligarchical, with sharp distinctions between nobility, commoners and slaves, but their government was a sort of primitive republic. There were four sections, each with a ruling lord who was the supreme power within his own boundaries. But the country as a whole was governed by a senate composed of the four lords and their most important followers.

The Tlaxcalan senate received Cortés's couriers and listened to their message courteously. One of the four principal lords, Maxixca, thought the strangers must be gods and that the Tlaxcalans would be wise to welcome them and seek an alliance against the Aztecs. An older lord, Xicotencatl — he was said to be one hundred and forty years old and was both blind and weak from age — opposed any such move. From what he had heard he judged the strangers to be more monsters than gods, but whatever they were there were not many of them and it would be a sign of weakness to allow them to invade Tlaxcala. If they were human they could be destroyed easily. If they were superhuman perhaps peace could be made with them later; but they should be tested and opposed. His view prevailed. The old man's son, also named Xicotencatl, was the Tlaxcalans' warrior chief and was charged with dealing with the invaders.

As a result Cortés and his column were met with hostility soon after they crossed the Tlaxcalan frontier — and before they were through they had been involved in the fiercest fighting any Spaniards had yet encountered in the New World. The first group of Tlaxcalan warriors was small, but they showed no fear of the Spaniards' horses or guns. Most of them were armed with the *macatl,* a broad, two-handed wooden sword edged with shards of volcanic glass. While Spanish armor would shatter these glass edges, the swordsmen could do dreadful damage on unarmored flesh — and in the first skirmish two of the Spanish horses were killed. The Spaniards managed to drive back the Tlaxcalans and the horses were immediately buried so that the Indians could not examine them and learn that they were not supernatural creatures. Four Spaniards had been wounded and at least seventeen Indians had been killed. The Spaniards spent a miserable night. For food they had only a few small hairless dogs found in a deserted village and the prickly, bulbous fruit of a cactus found growing everywhere. They cut the fat from a dead Indian and rendered it into oil to treat their wounds.

The next day two of the Totonacs that Cortés had sent to Tlaxcala as peace envoys returned to the Spaniards' camp, bringing word that a much larger Tlaxcalan force was approaching. With his interpreters Cortés rode out to meet the Tlaxcalans and, with the aid of Marina and Aguilar, delivered the *requerimiento* and followed it with a speech — which he had duly recorded by the notary, Diego de Godoy. He said that they had not come to make war and wanted only to pass in peace through Tlaxcalan territory.

Tlaxcalans responded with shouts and whistles and increased tempo on their wooden drums. Finally Cortés gave the order to charge. The thousand or so Tlaxcalans retreated in good order with the Spaniards in pursuit. They withdrew into a ravine through which the Spaniards had difficulty following. It was an ambush. Suddenly on both sides a much larger army appeared, shouting, screaming, waving flags with the standard of Xicotencatl, a white heron with outspread wings on a red ground. Stones from slings and swarms of arrows darkened the air. It was said that some of the archers could launch two, even three arrows at a time.

Finally the Spaniards made their way through the ravine and forced the action onto level ground where there was room for artillery and cavalry to operate. Pedro Morón, riding a mare borrowed from Juan de Sedeño, who had been wounded the day before, had his mount killed, an Indian swordsman decapitating it with one mighty blow of his *macatl*. Morón was mortally wounded, and this time the Indians took charge of the fallen horse and cut it up for distribution in all parts of Tlaxcala to demonstrate that it was not a magical creature.

By sunset the Indian masses had retreated. At various times Cortés later estimated that the attacking force had been from 100,000 to 139,000 strong, while Bernal Díaz del Castillo, usually far more realistic in such matters, said it was about 40,000. Whatever the size was, the Indians had sustained frightful casualties and so had the Spaniards in proportion to their numbers. Cortés withdrew his forces to the village of Yzompachtepetl, where he made his headquarters in a tower-like temple. The next day he raided some nearby villages seeking intelligence, and he dispatched another letter offering peace to the Tlaxcalans. Xicotencatl sent back a belligerent reply. That night the friar, Olmedo, and his chaplain, Juan Díaz, were up the entire night hearing confessions from Spaniards who were certain they would die.

In a gesture of sheer bravado, Xicotencatl the next day sent a train of porters carrying three hundred turkeys and two hundred baskets of tortillas. With the food came an insulting message: the strangers should eat and enjoy it, for when he defeated them in battle he did not want them to blame their humiliation on weakness from hunger.

Cortés rallied his men with a speech and gave practical instructions: the little company must stay tightly together with no straggling; the swordsmen must thrust rather than slash; horsemen were to ride three abreast and keep their lances well elevated, aiming for the eyes and face rather than the body; the musketeers and crossbowmen must stagger their fire, one shooting while another loaded.

The battle that followed, while not the last, was perhaps the decisive one. The Indians were finally driven from the field, but the Spaniards had many dead and wounded, and their spirits were sagging. "There was amongst us not one," Cortés later wrote to the king, "who was not very much afraid, seeing how deep into this country we were and among so many hostile people and so entirely without hope of help from anywhere." Cortés himself was ill with fever. Trying to break it, he had taken a strong dose of purgative that he had brought from Cuba. Before the medicine had worked he was back in the saddle and spent an entire day fighting, to the astonishment and admiration of his companions. "That he was made of iron is more than I would entirely agree with," wrote Bernal Díaz del Castillo in later years. "It is enough to say he made a good captain."

The Tlaxcalans sent to the Spanish camp five slaves and a stock of food, with a message: "If you are gods who eat flesh and blood, eat these Indians. If you are benign gods we offer you colored feathers and incense. If you are men, here are fowl and maize and cherries to eat."

A little later fifty Tlaxcalans arrived with porters bearing additional supplies of food. Cortés welcomed them to his camp and assured them that he and his followers were not gods but mortal men. The visitors wandered about the camp, displaying great curiosity about all they saw. One of the Totonac chiefs warned Cortés that they were spies. Drawing one of the visitors aside Cortés questioned him in a threatening manner and got a confession that they were indeed on a spying mission. Cortés ordered all the visitors seized and had their hands cut off, after which he sent them back to Tlaxcala.

The spying incident was followed by a night attack — an almost unheard-of thing among the natives, who usually stopped fighting at sundown. Some Tlaxcalan priests, however, had theorized that while the Spaniards were not gods, they were in truth children of the sun and that they lost their strength when the sun went down. It is more likely that Xicotencatl, a brave and pragmatic man, reasoned that in darkness his troops could not see the shining steel, the horses and the cannon of the Spaniards and would not fear them. The attack was brief, furious and confused. The Tlaxcalans took heavy losses. Afterward they were said to have sacrificed the priests who had urged on them this unconventional tactic.

The Spaniards, although they had managed to overcome the Indians each time, were more and more disheartened. Fifty-five of their companions were dead and almost all the survivors were wounded. There was much talk of retreating to the coast, where they could at least recover and possibly be reinforced by the Veracruz garrison.

Cortés spoke again to his men, both chiding and joking: "One must never turn one's back upon the enemy lest it appear to be a retreat. There is no retreat . . . which does not bring an infinity of woes to those who make it, to wit: shame, hunger, loss of friends, goods and arms, and death, which is the worst of them, but not the last, for the infamy endures forever." And then, half jokingly, he reminded them that while the rewards are great a soldier's life is hard, and he offered them some homely Spanish sayings to reinforce the point: "Where goes the ox that he will not draw the plow? Did you think, perhaps, that in some other place you would find fewer enemies? You are seeking a cat with five feet. . . . Wherever we go we shall find three leagues of bad road. . . ."

The sentiment for retreat diminished, and almost at the same time there was a change in fortunes. Moctezuma had followed the little war in Tlaxcala with intense interest. At first it had appeared that whatever the outcome it would be to his benefit: either the Spaniards would destroy the Tlaxcalans, an old and unyielding enemy, or the Tlaxcalans would destroy the Spaniards, which would solve an increasingly vexing problem. But now it appeared that neither side would triumph and that there might be a peaceful settlement and, worse, an alliance. He sent six nobles and a column of porters bearing gifts — including a thousand pesos' worth of gold — with a message for Cortés. The Spaniards must make no attempt to approach Mexico; the roads were too difficult and dangerous. He also inquired what annual tribute he would have to pay in gold, slaves and other goods if he, Moctezuma, acknowledged the superior authority of the Spanish king, a step he was ready to take if Cortés and his men would withdraw. Cortés made no commitment. While the parley was going on the Tlaxcalans made one final attack. The Spaniards repelled it easily — a fact that was not lost on the Aztec visitors.

The next day Xicotencatl seriously sued for peace. He brought a peace offering of gifts, a paltry one by comparison with that of Moctezuma, and Xicotencatl apologized that his nation was poor in everything except their love of independence. Cortés received him cordially and with respect and was invited to come to the city of Tlaxcala at once.

Cortés delayed leaving his campsite, however, anticipating another message from Moctezuma. He continued his discussions with the Aztec emissaries, who made every effort to cast doubt and suspicion on the Tlaxcalans. Cortés was to note later, in a letter to the king, the "discord and animosity" that existed between the Aztecs and the Tlaxcalans: "I was not a little pleased . . . it seemed to further my purpose considerably; consequently I might have the opportunity of subduing them more quickly. . . . So I maneuvered one against the other and thanked each side for their warnings and told each that I held his friendship to be of more worth than the other's."

Cortés's "design" was now beginning to be clear. He had known for months that there was much unrest in the Aztec empire and that it probably could be put to use. But until he had met the fierce fighting Tlaxcalans in a series of bitterly contested battles he had no certain idea just what a powerful weapon this animosity might be. Unassisted, the few Spaniards could probably do little against Moctezuma. In a hundred years of fighting, the Tlaxcalans had managed to preserve their independence, but they had never been able to crush the Aztec tyrants. Together the Spaniards and Tlaxcalans and other Aztec-hating Indians could probably bring it off.

Another Aztec emissary came from Moctezuma to the Spanish camp, this time with three times as much gold as before and a vast assortment of goods that were not so readily negotiable — fine textiles and beautiful featherwork. In a message

Moctezuma once more urged the Spaniards to return whence they had come.

Cortés accepted the gifts and ignored the advice. With his men in tight military formation and his various Indian allies straggling along behind, he marched into the city of Tlaxcala, accompanied by a multitude of Tlaxcalans, including even the ancient and blind Xicotencatl, who had originally rallied opposition to the foreigners. Now the old man repeatedly and affectionately addressed Cortés as Malintzin, Malinche's lord, a name that was to be adopted by virtually all the Indian peoples of Mexico.

It was a triumphal procession. The Spaniards were pelted with flowers and greeted with glad cries by crowds of people who had come from all parts of the little republic to witness the event. Several large buildings were prepared as dwellings for the Spaniards, with ample food and clean pallets. Cortés had persuaded the Aztec envoys to accompany him and kept them with him in his headquarters, knowing that they would report in detail to Moctezuma the cordial rapport between his two enemies.

Cortés estimated the population of Tlaxcala at one hundred fifty thousand families. In describing Tlaxcala in a letter to the king, he compared it with Granada, the fall of which in 1492 had been the most glorious victory in Spanish history. "The city is much larger than Granada," he wrote, "and very much stronger, with as good buildings and many more people than Granada had when it was taken, and very much better supplied with the produce of the land, namely bread, fowl and game and fresh-water fish and vegetables and other things they eat which are very good. There is in this city a market where each and every day upward of thirty thousand people come to buy and sell, without counting the other trade which goes on elsewhere in the city. In this market there is everything they might need or wish to trade; provisions as well as clothing and footwear. There is jewelry of gold and silver and precious stones and other ornaments of featherwork and all as well laid out as in any square or marketplace in the world. There is much pottery of many sorts and as good as the best in Spain. They sell a great deal of firewood and charcoal and medicinal and cooking herbs. There are establishments like barbers' where they have their hair washed and are shaved, and there are baths. Lastly there is amongst them every consequence of good order and courtesy, and they are such an orderly and intelligent people. . . ."

Cortés had reason to admire the security system. Someone stole a parcel of Spanish gold. When Cortés reported the fact to his hosts, the man was pursued to a nearby city, captured, brought back, made to give the gold back to the Spaniards and then clubbed to death in a public square. He had less evidence to support his claims of abundance. Tlaxcala was, as always, chronically short of cotton goods, salt and cacao, thanks to the blockade of the Aztecs, and they had very little in the way of gold, silver and jewels. As to whether Tlaxcala and its capture were in any way comparable to Granada and its fall, Charles V would not know; Granada had fallen before he was born.

It had taken Cortés three weeks or more of hard fighting to overcome the Tlaxcalans and to be welcomed into their capital; in the almost equal time he spent there as their guests, the Tlaxcalans did anything — or almost everything — in their power to make recompense. They heaped gifts on the Spaniards, and the Spaniards responded by giving the Tlaxcalans much of the cotton finery and featherwork they had received from the Aztecs, nonnegotiable goods for which they had little use but which their hosts regarded as the greatest luxuries. Five young girls, among them daughters of the principal lords, were presented to the Spaniards. Cortés tried to establish conditions for acceptance of the lovely creatures. He insisted that the idols be destroyed, human sacrifice stopped and the temples cleansed of blood and that Christianity be adopted. The Tlaxcalans bristled at the idea, and once more the prudent friar, Olmedo, persuaded Cortés to proceed more suavely.

It was agreed that the girls must be baptized before the Spaniards could accept them. Each of the girls had a substantial dowry. Xicotencatl had intended his daughter, Tecuilhuatzin, for Cortés. Cortés explained that he was already married (in addition, he was enjoying all the various talents of Marina). So the girl, renamed Luisa, was given to Pedro de Alvarado, the captain of the red hair and fiery temperament, whom the Tlaxcalans called Tonatiuh, the sun. Other girls, all with new Christian names, went to Juan Velázquez de León, Gonzalo de Sandoval, Cristóbal de Olid and Alonso Dávila. In addition, while the Tlaxcalans declined to abandon their old faith, they permitted the Spaniards to install a chapel in Xicotencatl's palace, to erect a cross at the point where Cortés had entered the city, and to cleanse and convert one of their many temples into a Christian shrine.

The Tlaxcalans told Cortés what they knew of their own history. They retold the legend of a fair-skinned god coming from the East. There had once been giants in their land, and they exhibited a thighbone as large as a normal man. Cortés politely asked if he might have it to send to his king. He was more attentive to their information about neighboring provinces and about Moctezuma and his realm. Nearby was the province of Huejotzingo, which, like Tlaxcala, was a republic and had long resisted Aztec tyranny; it was allied with Tlaxcala. Its men had fought side by side with the Tlaxcalans against Cortés. Cholula, twenty miles to the south, was a holy city, completely under Moctezuma's domination. Its people, while cowards on the battlefield, were crafty and treacherous, not to be trusted. Moctezuma could easily field an army of at least one hundred fifty thousand men. This was a considerable reduction from the three million Olintetl had mentioned, but it was still an awesome figure.

Great caution was urged on Cortés by Teuch, one of the Cempoala chieftains who had accompanied him from the coast. "Do not trouble yourself going on from here," he said. "As a youth I went to Mexico, and am experienced in the wars. I know that you and your companions are men, not gods; that you hunger and thirst and weary as men do. . . . Beyond this province lie so many people that one hundred thousand men will fight you now, and when these are dead or vanquished that many again will come forward, and again and again by the hundred thousand, and you and yours, though you be invincible, will die wearied of fighting. For as I have told you, I know you are men, and all I can say is that you should think carefully about what I have said. But if you determine to die, then I shall go with you."

Moctezuma too had been gathering intelligence — from his emissaries who were observing the Spaniards in Tlaxcala, and from spies who kept a close watch on the Spaniards left behind in Veracruz. He still considered the omens, the reports of the Spaniards' invincibility in the Tlaxcalan battles and the ancient predictions. He called into consultation priests and astrologers and he engaged in long periods of meditation in a darkened room. The result was that he changed his mind once more. He gathered a still larger gift of gold and sent it with a message to Cortés in Tlaxcala. The Spaniards might come to the Aztec capital and visit with him, and they should come by way of Cholula, where arrangements were being made to receive them in appropriate style.

The Tlaxcalans warned Cortés of a trap. They pointed out that while representatives of far distant provinces had come to pay homage and respect in Tlaxcala, no one had come from Cholula, which was less than twenty miles away. Cortés thereupon sent a message to Cholula demanding that an embassy be sent to him. The Cholulans then sent representatives to confer with him. The Tlaxcalans pointed out that these emissaries were neither chiefs nor nobles but persons of inferior rank, an insult to the Spaniards. Cortés sent another message in which the substance of the much-used *requerimiento* was embodied; Cholula's chieftains must come to Tlaxcala and submit to Cortés as surrogate for the king of Spain — otherwise they would be considered rebels. The chiefs from Cholula obeyed the order — which must have seemed very strange to them — and arrived the next day, blaming their delay on their distrust of the Tlaxcalans.

So, on October 11, after twenty days of rest, Cortés and his little army, now mostly recovered from their wounds, well fed and rested, their weapons clean and shining, set out on the penultimate stage of their march to Mexico. The Tlaxcalans deplored their leaving and still argued that the advance should be made by way of friendly, dependable Huejotzingo. Maxixca, who had become devotedly attached to Cortés, seldom letting the Spanish captain out of his sight, wept at his departure. And an army of one hundred thousand Tlaxcalan warriors followed the Spaniards on their march, ready for anything the crafty Cholulans and the hated Aztecs might have waiting for them. Cortés, used to marching with a few hundred of his own men and a few thousand Indian allies, thought such an awesome array ill-advised. He persuaded most of them to turn back, accepting the services of only five thousand or so. And when, after an easy march, they came to Cholula, he asked the Tlaxcalans to camp outside the city while and he and his fellow Spaniards went in alone.

EIGHT

Cholula

CHOLULA WAS A city of some twenty thousand houses and, the
Spaniards said, as many temples as there were days in the year,
for it was a holy city, sacred to the worship of its supposed
founder, the kindly old god Quetzalcoatl. Pilgrims came from
all parts of Mexico to worship here in the little city nestled on
the plateau just below and to the east of the giant snow-
capped volcanoes, Popocatepetl and Ixtaccihuatl. The land
was fertile and well watered by streams flowing down from the
snowpack on the mountains.

The main temple of Quetzalcoatl was atop a giant pyramid,
the largest such structure in a land of pyramids, its base
covering twenty-four acres. Traditionally the only sacrifices
here had been those of small birds, animals and reptiles, in
keeping with the Quetzalcoatl tradition. In the years before
the arrival of the Spaniards, however, there were occasional
human sacrifices. This was, perhaps, the result of the domination of immolation-mad Aztecs. For many years Cholula had
been independent, its affairs administered by a council of
nobles, its army commanded by a general selected by the
priests. It had been allied with the similarly independent city-
states of Tlaxcala and Huejotzingo. But gradually it had

fallen under the control of the Aztecs, and the old alliance had been broken.

The Spaniards, when they first saw it, were enchanted with the landscape that was much like their homeland. The people, too, were handsome, dressed in long cotton robes. Cortés was later to write of Cholula: "The city itself is more beautiful to look at than any in Spain, for it is very well proportioned and has many towers. . . . From here to the coast I have seen no city so fit for Spaniards to live in, for it has water and some common lands suitable for raising cattle, which none of those we previously saw had, for there are so many people living in these parts that not one foot of land is uncultivated, and yet in many places they suffer hardships for lack of bread. And there are many poor people who beg from the rich in the streets as the poor do in Spain and in other civilized places."

The march from Tlaxcala to Cholula could have been made in a day, but Cortés halted and camped in the dry bed of the Atoyac River before entering Cholula. Here he was visited by Cholulan priests and dignitaries bearing food and gifts. The next day the Spaniards, marching in single file, were escorted into the city. Priests attended them along the way, singing, blowing on bone flutes and shell trumpets, beating drums, carrying draped idols and swinging clay censers with smoking copal incense. Bouquets and garlands of flowers were thrown. The Tlaxcalans remained behind, camped outside the city on Cortés's orders.

The Spaniards were installed in a large building with several courts, an adjunct of one of the temples. Each man was given a turkey and a supply of tortillas, vegetables and fruit.

Despite this seemingly cordial beginning Cholula would soon be the scene of the bloodiest massacre the Spaniards had yet perpetrated. Cortés subsequently reported that three thousand Indians were killed there in a few hours (although ten years later his agent, responding to a royal inquiry, admitted only that "some" had been killed). Other estimates of the casualties ran from six thousand to twenty thousand. Fray

Bartolomé de las Casas reflected on it bitterly: "The Spaniards decided to make here a massacre or punishment in order to disseminate fear . . . a cruel and outstanding massacre so that these poor tame sheep should tremble." The Cholula incident was a dreadful lapse from the ethical Christianity Cortés so passionately professed. Centuries of later argument failed to solve the question of whether it was deliberate terrorism or tragically harsh response to the exigencies of the moment.

The Cholulans' initial hospitality to the Spaniards came to an end within four days of the newcomers' arrival. Food that the natives at first brought to the visitors in abundance dwindled. And there were, according to the Spaniards, indications of approaching danger. A group of Aztec emissaries who had been in Cholula left for Mexico, taking with them the principal Aztec ambassador, who had been with Cortés since before the fall of Tlaxcala. Cortés also was informed that the direct road to Mexico, running through the pass between the two great volcanoes, was blocked with logs and boulders. The alternate road, winding around the south side of Popocatepetl, was a dangerous one, cut by barrancas and canyons. Many streets in Cholula were barricaded. Large piles of rocks were seen on the flat roofs of buildings, ready to be thrown down on passersby.

And the Spaniards' Totonac allies reported many more alarming things. Holes had been dug in the streets, filled with sharpened stakes and covered with brush as traps for the Spaniards' horses. Huge cooking pots with piles of tomatoes and peppers were being made ready for a cannibalistic feast. Small children were being sacrificed to the god of war. Other more fortunate children and the women were being evacuated. And a large Aztec army was camped just outside the city, ready to sweep in and overcome the foreigners. All the foreigners were to be captured alive. Twenty were to be left for the Cholulans to sacrifice and eat; the remainder were to be taken to Mexico for sacrifice.

Cortés pondered these rumors. He had chosen to come by

way of Cholula because Moctezuma's ambassadors had suggested it rather than the more direct route through Huejotzingo, whose people hated the Aztecs just as much as did the Tlaxcalans. Strategically it made sense. With Cholula in his hands it would be easier to safeguard communication with the Gulf coast, his supply line if he ever received supplies, or his retreat route if it came to that. His exchanges with Moctezuma were baffling. In one message that irresolute monarch seemed to be inviting him to come to Mexico; in the next he would command Cortés to return whence he had come.

Marina came to Cortés with a frightening story. She had been approached by a Cholulan woman who admired her and wanted her to remain in Cholula and marry one of her sons. She was urged to accept in order to save her own life, because the Spaniards were soon to be attacked. The Cholulan woman knew because her husband, a chief, had been given a golden drum by the Aztecs as an inducement to take part in the assault. Marina pretended to agree but begged for time, saying she would have to find someone to carry her personal goods, her clothes and her jewelry, and she persuaded the older woman to wait in her quarters while she did so.

When Cortés heard Marina's story he seized a priest and, through Marina's interpretation, forced a confession out of him that confirmed the story. He next called his principal men together and explained what he knew. Some were in favor of an immediate retreat, but Cortés opposed it. They must smite or be smitten, he insisted, and finally his followers agreed. The principal lords of Cholula were then called to the Spaniards' quarters. Cortés, through Marina, told them he realized they were unwelcome in Cholula and would withdraw on the following morning if the Cholulans would only provide him with several thousand porters. They agreed, although why the Spaniards should need so many more porters than they had brought with them must have been puzzling. Cortés next called the Aztec emissaries and told them of the supposed plot, saying the Cholulans had informed him they were acting on Moctezuma's orders. The Aztecs vigorously denied it. Cortés

appeared to accept their denial and declared that the Cholulans must be punished for their treachery and their effort to
shift the blame to Moctezuma; it was an affront to the great
man, which his faithful friend, Cortés, resented. The Aztecs
were then locked up for the night to avoid any leak of Cortés's
plan for retribution. Cortés also sent word to the Tlaxcalan
allies camped outside the city, to be ready for action.

At daybreak several thousand Cholulans swarmed into the
courtyards of the Spaniards' quarters. Some Spaniards later
pointed out they did not look like porters; many of them
carried weapons. When all were inside, Cortés ordered the
three doorways closed and barred. He took the commanding
chiefs into closed rooms and questioned them one by one.
They admitted plans for attack but placed the blame on
Moctezuma. Cortés then mounted his horse. He had a musket
fired as a signal and the general assault began. Spanish artillery, in place and primed, cut bloody swaths through the
natives packed in the courtyards; Spanish lances, crossbows
and swords finished the work. The Tlaxcalans outside the city,
having heard the musket fired, stormed into the streets, killing, burning, looting. Buildings and streets were soon piled
deep with corpses. Some Cholulans retreated to the top of the
great pyramid. It was said that Quetzalcoatl had left behind a
legend: if his city was ever threatened certain stones should be
pried loose from the pyramid. A torrent of water would gush
out and destroy the invaders, quench any fire. The indicated
stones were removed and nothing happened. The Cholulans
threw themselves to death down the face of the pyramid.

While there is no tangible reason to doubt the sincerity of
Cortés's conviction that a plot existed, the severity of the
retaliation was beyond anything Cortés had done before, even
when he and his men were in far graver danger. Years later,
however, López de Gómara, who was not present but learned
about it from Cortés, was to write: "Truth to tell, it is war and
warriors that really persuade the Indians to give up their
idols, their bestial rites, and their abominable bloody sacrifices
and the eating of men, which is directly contrary to the laws of

God and nature, and it is thus that of their own free will and consent they more quickly receive, listen to and believe our preachers and accept the gospel and baptism, which is what Christianity and faith consist of."

The surviving Cholulans were put to work disposing of the bodies and cleaning up the city. The women and children who had fled the city returned. And from that point on, Cortés insisted, Cholula was a model of loyalty and obedience to the Spanish king and the tenets of Christianity.

The bulk of the Tlaxcalans — all except one thousand who were to continue to serve as porters — were sent home with vast quantities of loot, including salt and cotton, which they badly needed. The smaller group of Totonac warriors and porters were sent back to Cempoala with more loot and with messages for Juan de Escalante at Veracruz.

The puzzling exchange of messages with Moctezuma went on and on. Each message from the Aztec ruler was accompanied with gifts of gold, worth one thousand pesos one time, three thousand the next, ten thousand the next. The rapidity with which Moctezuma changed his mind as to whether the Spaniards should proceed to Mexico, stay where they were or withdraw was indicative of the unsettled state of his mind. But for the Spaniards the increasing lavishness of Moctezuma's gifts of gold resolved any doubt there might have been: they had to go where it came from.

In all the Spaniards spent twenty days in Cholula, preparing for what they hoped would be the final stage in their march of conquest. One of the remarkable by-products of the stay was an exploration of the volcano Popocatepetl, which was in eruption at the time, its summit belching smoke and ashes. Diego de Ordaz was detailed to attempt to climb to the peak. Leaving an Indian escort at a lower level, beyond which they were afraid to go, Ordaz and nine companions ascended as far as they could. Bernal Díaz said they reached the crater at the summit and saw the fiery pit from which all the disturbance came. Others maintained that they were prevented from reaching the summit by deep snow, high winds and the storm of hot cinders and lava. Later Ordaz, when he received a coat of arms, had the smoking volcano as one of its emblems. And Cortés was later to take advantage of the knowledge Ordaz had gained. When he began the manufacture of gunpowder in Mexico he sent men up the volcano to collect sulfur.

Ordaz and his men made another discovery. In the pass between the mountains they found a good road leading to the west. From this vantage point they could see the great lake-bound city of Tenochtitlan, the capital of Mexico, glimmering in the distance.

For Cortés it was valuable intelligence. He confronted Moctezuma's emissaries with his knowledge of the existence of a good and more direct road to Mexico — instead of the circuitous and dangerous one they had urged him to take. They conceded that this was correct, but the good road went close to Huejotzingo, whose people were Moctezuma's enemies. However, they agreed to accompany him on this road and would have supplies sent out from the capital to sustain the Spaniards on their march.

On November 1 the Spaniards left Cholula, which Cortés declared to be perfectly subdued. In two days of marching they reached the pass between the two high mountains. A little farther, they found newly erected straw huts, put up for the Spaniards' convenience, and a supply of firewood to keep them warm at that high elevation. Here Cortés was visited by another delegation from the Aztec ruler, headed by a man said to be Moctezuma's brother. With them came the usual gift of gold and another plea that the strangers make no attempt to visit the capital: provisions were short; Moctezuma was ill and unable to travel; the road was dangerous; and, since much of the city was under water, it could be approached only by canoes. "I received them very well," Cortés later wrote to his king, "and gave them some of the things from Spain which they hold in great esteem. . . . [And to their insistence that I advance no farther but return whence I had come] I replied that were it in my power to return I would do so to please

Moctezuma, but that I had come to this land by Your Majesty's commands, and that the principal thing of which I had been ordered to give an account was Moctezuma and his great city, of which and of whom Your Majesty had known for many years." And he went on: "I told him to beg Moctezuma on my behalf to acquiesce in my journey, because no harm would come of it to his person or his land, rather it would be to his advantage; and that once I had seen him, should he still not wish me to remain in his company, I would then return. . . ."

Cortés also warned his visitors that the Spaniards never slept at night and would kill anyone who approached their camp after sundown. That very night fifteen natives were killed for alleged spying on the Spaniards, although it is probable that they were trying to do nothing more than satisfy their curiosity about these strange creatures.

Farther on, Cortés was met by a delegation headed by Cacama, the young man Moctezuma had installed as king of Texcoco (it was his disappointed brother, Ixtlilxochitl, who had offered an alliance "against tyranny" to Cortés in Veracruz). He came as a personal representative of his uncle, Moctezuma, and was elaborately attired and in splendid style; he rode in a litter decorated with jewels and colorful plumes and attended by a richly dressed retinue and a considerable force of less important Indians. Once more the plea was made that Cortés advance no farther. And there was the suggestion that if they insisted on doing so Cacama and his men would physically block the way. Cortés was polite to the young king, offered the usual gifts of beads and trinkets, and marched on, paying no attention to the warning.

Soon they were on one of the interconnecting causeways across the lake which would lead them to Tenochtitlan. They marveled at the *chinampas,* the floating gardens on which the people grew abundant crops of vegetables, fruits and colorful flowers and shrubs, and paused in the island city of Ixtapalapa, only five miles south-southeast of the heart of Tenochtitlan. The local chief, Cuitlahuac, although he had counseled his brother, Moctezuma, against permitting the Spaniards to approach, was courteous, hospitable and generous. He pressed gifts of gold, featherwork and female slaves on the foreigners.

Later, in a dispatch, Cortés recalled Ixtapalapa in lyrical terms. It had, he said, from twelve thousand to fifteen thousand inhabitants. "It is built by the side of a great salt lake, half of it on the water and the other half on dry land. The chief of this city has some new houses which, although as yet unfinished, are as good as the best in Spain; that is, in respect of size and workmanship both in their masonry and woodwork and their floors, and furnishings for every sort of household task. . . . They have many upper and lower rooms and cool gardens with many trees and sweet-smelling flowers; likewise there are pools of fresh water, very well made and with steps leading down to the bottom. There is a very large kitchen garden next to the house and overlooking it a gallery with beautiful corridors and rooms, and in the garden a large reservoir of fresh water, well built with fine stonework, around which runs a well-tiled pavement so wide that four people can walk there abreast. It is 400 paces square, which is 1,600 paces around the edge. Beyond the pavement, toward the wall of the garden, there is a latticework of canes, behind which are all manner of shrubs and scented herbs. Within the pool are many fish and birds, wild ducks and widgeons . . . the water is often almost covered with them."

To the battle-soiled, travel-weary Spaniards it was much richer and more luxurious than anything they had known before, and a longer stay must have been very tempting. But it also was pregnant with promise; if Ixtapalapa, a secondary city, was this rich, this elegant, then the capital, whose shining buildings and awesome temples could be seen clearly in the sparkling air of the highlands, must be the ultimate in an adventurer's dreams.

And so they did not tarry. They rested overnight and in the morning the column set out, marching briskly along a causeway toward the end of the rainbow.

NINE

Tenochtitlan (II)

It was Tuesday, November 8, 1519, still less than a year since the start of the adventure. The little Spanish army, followed by a much larger army of native allies, set out on the causeway leading to Mexico. The causeway, built of earth and stone, was wide enough so that eight horsemen could ride abreast. The foot soldiers marched in close order, "beard on shoulder," as Bernal Díaz put it. And he added: "As we are men and feared death we never ceased thinking about it, and we marched by small stages, commending ourselves to God and to His blessed Mother Our Lady." Some of the Spaniards wore traditional iron helmets, breastplates and leg guards, and they clanked as they moved. Most of them, however, allowed Indian porters to carry this hardware and wore, instead, caps and protective coats made of thick quilted cotton. It was a warm day, and both horses and men sweated profusely. Their weapons, swords, lances, crossbows and muskets, were generally cleaner and better kept than the men themselves, and they glinted in the fall sun.

The causeway from Ixtapalapa, running toward the north, was after a time joined by another one running toward the northeast from Coyoacan. Here at the junction there was a small fortress with gates on either side, controlling access to the main causeway leading to Tenochtitlan. On the east side of the causeway the water was saline, and here and there were evaporative flats where salt was collected. On the west side the water was sweet. At intervals there were breaks in the causeway, bridged over for foot traffic and with gates to control the flow of water. On both lakes there were many canoes, carrying produce and trade goods to and from the great city. The boatmen paddled their craft close to the causeway for a look at the strange sight: four-legged animals that the natives took to be giant deer, capable of carrying men on their backs; and the equally exotic two-legged animals with pale skin, bushy beards and outlandish clothing.

Later a Franciscan friar, Bernardino de Sahagún, took down from the natives their recollections of that fateful scene — as strange as a visitation of extraterrestrial creatures might be in the modern world: "Four stags [horses] come in front, the first like leaders. . . . They prance, turn, look backward, then to one side and the other and then to all sides. . . . Then the dogs pass along, their noses to the ground. They follow the footsteps, panting heavily. At the front there is only the banner. . . . The bearer carries it on his shoulders. He shakes it and makes it move in circles, then from one side to the other. Following him are those with unsheathed iron swords, the swords shining and glittering. They carry their shields on their shoulders. . . . The second group is made up of horses with riders on their backs. These wear armor of cotton, their shields covered with leather, and their iron swords which they allow to hang from the trappings of the horses. They come with bells. They almost clang, the horses neigh, they sweat much, water almost runs down their sides. The foam from their mouths drips on the ground. . . . In running they make a great clattering; they make a noise as if someone were throwing stones. The instant they raise their foot they stir up the ground. . . . The third file is made up of crossbowmen. . . .

They . . . carry . . . [the weapons] completely full . . . of iron arrows. . . . Their heads [are] enveloped in cotton armor, and above [they] have put on quetzal feathers which stick out on all sides." Then came the musketeers, who carried their weapons on their shoulders or horizontally. And finally Cortés's Indian allies.

If, as Bernal Díaz said, there was uneasiness among the Spaniards, there was at least as much in the Aztec delegation which was moving out of the city to meet them. Borne aloft in a litter decorated with jewels and luxurious draperies and carried by a group of dignified noblemen, richly dressed except for their bare feet, came Moctezuma. The king sat under a canopy decorated with green feathers. At his sides were his nephew, Cacama, king of Texcoco, and his brother, Cuitlahuac, both of whom had already met Cortés — Cuitlahuac only the day before. It would give Moctezuma a certain advantage in being able to identify Cortés easily. There was also a political element in the choice of these two: both had strenuously opposed permitting the Spaniards to enter the city. Moctezuma's attitude toward the strangers had veered widely, from distrust to fear to a fatalistic acceptance of the inevitable. He had, recently, examined the heads of two Spaniards killed in a skirmish near Veracruz and brought to him by swift runners. The heads were strange, thickly bearded, but they also seemed human rather than godlike. So, the Spaniards were probably as human as the Aztecs themselves; perhaps they were children or servants of a powerful god. Out of Moctezuma's nightmarish anxieties had come at least a tentative decision: tact and hospitality might be more effective than hostility in coping with an as yet ill-defined peril.

At the twin towers the Spaniards halted while nearly a thousand Aztecs ceremoniously greeted them by touching the earth and then kissing their hands, cotton mantles draped over copper-colored skins, rich plumes waving and bobbing on their headdresses. These preliminary ceremonies required almost an hour. Then the procession moved on through the

gates and onto the broad causeway leading into the city—
and to the advancing royal column. Finally the two groups
met. The barefoot noblemen swept the pavement and spread
out embroidered cotton cloths, always keeping their eyes on
the ground, never gazing at their monarch. When all was ready
Moctezuma placed his feet, shod in gold-soled gem-studded
sandals, on the carpeted pavement and with Cacama holding
his right elbow, Cuitlahuac his left, advanced to an encounter
that would shape both his own destiny and that of his nation.

Cortés swung himself off his horse, handed the reins to a
page and advanced, his eyes curious and watchful—a bold-
ness that may have appalled Moctezuma's attendants, who had
been trained from birth to avert their eyes from their ruler.
The two men halted a few paces apart, staring at each other.
Cortés approached, arms outspread, ready to clutch Moctezuma
to his chest in the traditional Spanish embrace of greeting.
Cacama and Cuitlahuac sprang forward and grasped the Spaniard's arms to prevent the sacrilege of physical contact with
their king. Then, stepping back, Cacama and Cuitlahuac
touched the earth and kissed their hands, and Moctezuma
immediately repeated the gesture of respect and welcome.

Through the tandem interpretation of Aguilar and Marina,
Cortés advised Moctezuma to be cheerful and unafraid, that
the Spaniards regarded him with affection and reverence.
With this he took from his chest a necklace of glass beads shot
through with gold and perfumed with musk, and placed it
over Moctezuma's head. Intrinsically it was insignificant compared with the finery the Aztec ruler already wore. Moctezuma
had servants bring forward two necklaces of red shells hung
with life-size shrimps made of gold. These he placed around
Cortés's neck. Then, turning, he took Cacama's hand and
walked toward the city, with Cortés and Cuitlahuac following,
also hand in hand.

The city streets, all swept clean, were curiously deserted,
although dark eyes occasionally could be seen staring out of
the interior gloom of the houses.

The causeway led the strange procession to a broad avenue and the heart of the city, where a great walled compound enclosed the principal temples. To the right of the compound was a vast, flat-roofed palace, which the Spaniards learned was Moctezuma's principal residence. To the left, only a few hundred yards away, was a similar structure, which had been the palace of Axayacatl, Moctezuma's father.

Axayacatl's palace was to be the Spaniards' home and garrison. It had for a time been used as a convent for young women devoting themselves to a religious life; it also, the Spaniards were to discover later, contained storerooms for the imperial treasure — a discovery which Moctezuma must have suspected the eager Spaniards would make, for, as one native chronicler put it, "they were greedy for gold, like pigs, like monkeys."

But for the moment it was nothing more than a vast structure with spotlessly clean rooms, halls, courtyards and corridors, more than enough space for all the Spaniards, their horses and their Indian allies. Walls, ceilings and floors were highly polished, some of them covered or draped with cotton or feather tapestries. Braziers burned in most of the rooms, throwing out a sweet-smelling smoke. Neatly arranged on the floors were pallets of woven straw for sleeping.

Taking Cortés by the hand, Moctezuma led him through the courtyard and into a great hall where he bade him sit on a stool ornamented with gold and gems, to eat and rest himself and to expect another visit from Moctezuma later in the day. Large quantities of food were brought in by an army of nervous servants, and the Spaniards made the most of it. Cortés, meanwhile, was busy assigning quarters to the various sectors of his entourage and overseeing emplacement of the artillery pieces on the roof.

The Franciscan friar, Sahagún, reconstructed the natives' view of this scene also: "After entering the great palaces in the royal city they discharged their muskets. On firing them it is seen how it thunders, how it lightens, how the smoke extends. It turns night, the sky is darkened by smoke, the smoke covers all the land, spreading all over the country until it smells of sulfur, which robs the senses and the consciousness."

The noise and stink of the Spaniards' gunfire apparently did not deter Moctezuma. He returned to Axayacatl's palace with gold and silver ornaments and bundles of fabrics as a gift for Cortés. And either that day or the following morning — sources differ — he made a speech of welcome which Cortés later reported in a dispatch to the king:

"For a long time we have known from the writings of our ancestors that neither I, nor any of those who dwell in this land, are natives of it, but foreigners who came from very distant parts; and likewise we know that a chieftain, one of whom they were all vassals, brought our people to this region. And he returned to his native land and after many years came again, by which time all those who had remained were married to native women and had built villages and raised children. And when he wished to lead them away again they would not go nor even admit him as their chief; and so he departed. And we have always held that those who descended from him would come and conquer this land and take us as their vassals. So, because of the place from which you claim to come, namely, from where the sun rises, and the things you tell us of the great lord or king who sent you here, we believe and are certain that he is our natural lord, especially as you say he has known of us for some time. So be assured that we shall obey you and hold you as our lord in place of that great sovereign of whom you speak. . . . And in all the land that lies in my domain, you may command as you will, for you shall be obeyed; and all that we own is for you to dispose of as you choose. Thus, as you are in your own country and your own house, rest now from the hardships of your journey and the battles which you have fought."

Moctezuma went on, Cortés said, to say that he knew that his enemies with whom Cortés had spoken had told lies of him, that the walls of his houses and the household goods were all of gold, and that he, Moctezuma, claimed to be divine. He pointed to the walls and floor of the room in which they were seated, explaining that they were of stone and lime, not gold. He raised his tunic to show Cortés that his body was of human flesh. Finally he said that it was true that he had some gold that had been left to him by his ancestors, and that any or all of it could be had by the Spaniards for the asking.

The accuracy of the statement might be questioned. It had progressed through two stages of translation. And Cortés, in his dispatches, was inclined to report what he wished. But one thing seemed clear. Moctezuma identified Cortés and the king he served with a leader the Aztecs had once rejected, and he was now willing to make amends for that rejection.

The next day Cortés, with a few of his picked captains and men, paid a formal call on Moctezuma in his own palace. The vast structure was similar in style to the palace in which the Spaniards were housed, but it was newer, more elegantly finished and more luxuriously furnished. The woodwork and ceilings were of finely carved cedar and other timber, cunningly joined together without nails. Floors and walls were of highly polished stone — porphyry, marble, jasper. A hundred or more huge rooms were clustered around the three great courtyards, in one of which a fountain played. Moctezuma met them in one of the courtyards and escorted them into an audience room, where he bade Cortés to take a seat at his side and the other Spaniards to be seated. The few courtiers who were on hand were dismayed by the way in which the Spaniards stared directly at their king, a thing none of them dared do. It seemed not to matter to Moctezuma; his air was one of open friendliness and courtesy.

Cortés, relying on his "tongues," Marina and Aguilar, launched into an evangelical harangue. Bernal Díaz, who was one of the soldiers who had come along, remembered him saying: ". . . that we were Christians and adored one true God, called Jesus Christ, who died to save us." Bernal Díaz went on, "He told how He was resurrected on the third day and was in heaven. . . . What they took for gods were not gods at all but devils, which were very bad things. Wherever we had raised crosses they dared not appear, as he would notice as time went on.

"Cortés also told about the creation of the world and about how all of us were brothers, children of Adam and Eve, and he said that it was wrong to adore idols and to sacrifice men and women because we were brothers. As time went on our lord and king would send men who led holy lives, better than ourselves, who would explain everything."

And Moctezuma, according to Bernal Díaz, replied: "Señor Malinche, I have understood what you have said . . . about three gods and the cross, and the other things you have spoken about in towns through which you have passed. We have not answered any of it, for here we have always worshipped our own gods and hold them to be good; so must yours be. For the present do not talk to us about them any more. This about the creation of the world we have believed for ages past. For this reason we are sure that you are those whom our ancestors predicted would come. . . ."

Moctezuma, said Bernal Díaz, referred to the voyages of Córdoba and Grijalva and asked if Cortés and his men were of the same people. Assured that they were, he said that he had considered honoring those earlier voyagers in his cities. If he had seemed reluctant to permit Cortés and his men to advance on his capital it was because most of his vassals feared the Spaniards, thinking them to be gods, and feared, too, the thunder and lightning they made with their guns, their horses. Now that he could see them in person he knew that they were not gods, but only brave men.

The Spaniards may have been surprised by Moctezuma's apparent tolerance of the existence of other gods, other beliefs. It was not the sort of tolerance the God-fearing Spaniards extended to others. But if Cortés was prepared to argue the point he did not do so, for once more Moctezuma presented his visitors with gifts: two gold necklaces and two *cargas* (approximately fifty pounds each) of finely woven cloth for each man. The Spaniards took their leave, politely raising their quilted cotton caps to their host.

Moctezuma was about forty years of age at the time, older than most of his visitors. He was taller than most of his countrymen, slender but with a muscular frame. His skin was a light cinnamon color, his hair a glossy black, worn long enough to cover his ears. His eyes were dark, his face long, and on his chin and at the corners of his mouth were a few thin strands of whiskers. His body and clothing were immaculate. He bathed daily and changed his clothing, a sort of fussiness that astonished the travel-tainted Spaniards. They noted with approval that while his opportunities for sexual pleasures were vast he was not lecherous, was discreet and conventional, not a practitioner of sodomy (the supposition that most Mexicans did practice it was one of the many myths created in the course of the conquest). His bearing was dignified and he was treated with both reverence and deference by his subjects.

Even the most powerful of Moctezuma's nobles, when coming to call, removed their sandals outside the palace and changed from fine garments to humble, simple ones. In the royal presence they stared only at the floor, never raising their eyes, and any discourse was preceded by three bows. When they left they backed away, eyes still on the floor.

His cooks prepared at least thirty kinds of food for each meal — turkey, pheasant, partridge, quail, pigeon, duck, venison, wild pig, vegetables and fruit. The foods were served over charcoal braziers made of clay; often the king did no more than sample a few of the dishes, leaving the remainder for his attendants. He ate from red and black Cholula pottery plates while sitting on a low bench by a low table, shielded from view by a screen worked with gold. A few counselors might attend while he ate, and Moctezuma sometimes offered them portions from the dishes he particularly liked. This they accepted reverently, arising to eat it. From gold cups Moctezuma drank foaming chocolate, whipped to a froth. It was a drink the Spaniards heard gave him special powers over women.

When he had finished eating, maidservants brought bowls of water, in which he washed his hands. After eating he would light a pipe filled with a mixture of fragrant resin and tobacco, and watch while court entertainers — singers, dancers, jugglers and jesters — performed. His household was a vast and varied one: servants and guards by the thousands, wives and concubines by the hundreds, gold- and silversmiths, featherworkers, gardeners, carpenters, weavers, woodcarvers. There were, in addition, aviaries, botanical and zoological gardens, fountains, baths — luxuries and opulence such as none of the Spaniards had ever known.

Nor was the city less amazing. On the fourth day after their arrival Cortés sent word to Moctezuma asking permission to visit the markets and the temples. Knowledge of both — the markets, which supported the physical life of the city, and the temples, which supported the spiritual — was essential to any logical planning of future strategy. Moctezuma agreed to the request. Certain of his courtiers were assigned to accompany Cortés and his men to the great *tianguiz*, or market, at Tlatelolco, a twin city situated on the same island but slightly to the north and west of Tenochtitlan. Cortés and his captains, mounted on horseback and followed by most of the foot soldiers, rode the few miles to Tlatelolco, while Moctezuma promised to meet them later at the great temple in Tenochtitlan.

Every fifth day was market day, and activity was at a peak when the Spaniards arrived at the scene. It was a catalog of the country's wealth, its skills, artistry, commerce and even of the social structure, for among the goods displayed for sale were slaves, yoked to long poles. There were vast quantities of food, maize and the ever-present tortillas, fruits from temperate highlands and tropical lowlands. There were oddities such as a cheese-flavored cake made from a scum gathered from the surface of the lake. And there was a sweet-tasting cactus fruit that would stain the urine blood-red, a cause for alarm among the Spaniards at first and much practical joking afterward.

Meat was offered, either butchered or alive and ready for killing — fowl, wild pigs, deer, small dogs castrated and fattened. Cotton was offered in a raw state or spun and woven — and there was even finer fiber made from rabbit fur. There were bales of skins, both untreated and tanned — rabbits and hares, ocelot, jaguar, deer. Apothecaries displayed large stocks of herbs, roots, powders and infusions. One Spaniard noted, perhaps with a sense of relief, "They . . . even have a special and well-known herb for killing lice." Some traders offered goose quills filled with gold dust or nuggets. Nearby goldsmiths manufactured and sold curious and beautiful ornaments, implements and toys made of gold including articulated human and animal figures and fish with alternate scales of gold and silver. Lapidaries cut and polished the green chalchihuitl, or jadeite, and other precious stones. Barbers cut hair and trimmed sparse beards with blades of obsidian. There was pottery in many shapes, colors and textures, hard woods carved into trays and bowls and idols, textiles ranging from the plain and utilitarian to intricate designs worked in embroidery or feathers. And there were piles of cacao beans, which served the double function of currency and the source for chocolate. Estimates of the number of people engaged in selling and buying ranged from forty thousand to eighty thousand. To settle the quarrels that abound in any marketplace in the world there were armed guards and a tribunal of judges.

Awed by the wealth, the abundance of goods and the efficient organization of the economy, the Spaniards retraced their way back to Tenochtitlan and the great temple, where they were to meet Moctezuma. The temple area, enclosed by a wall, contained forty or more individual temples, almost all supported by pyramidal stone structures, was paved with shining white stone, on which the Spaniards' horses slipped and slid. The principal temple, or *teocalli* (god house), towered over all the rest. The Spaniards bounded up the steep steps — one of them counted 114 steps — to the broad, flat summit. Here Moctezuma told Cortés that he should have waited for servants to carry him up, to which Cortés, breathing heavily, replied that he and his comrades never became weary, no matter how great the exertion. If Moctezuma was aware of the braggadocio he ignored it. Taking Cortés by the hand, he led him to the edge of the broad platform and, pointing this way and that, gave a sightseeing lecture.

Back to the north was Tlatelolco, where they had just been. Beyond it a causeway ran across the northern part of the lake to Tepeyac. To the west another causeway crossed the lake to Tacuba, with a branch running slightly to the south to the heights of Chapultepec (the hill of the grasshopper). Along this causeway was built a double-channeled aqueduct, which supplied the city with sweet water from springs at Chapultepec. To the south lay the causeway along which the Spaniards had approached the city. Directly east across the lake, but not connected by causeway, was Texcoco, an older city than Tenochtitlan and at least as large. Texcoco, Tacuba and Tenochtitlan formed the three-city alliance on which was based the Aztec empire — or more correctly federation, since most of the city-states had rulers of their own. Around the shores of the lake were at least fifty smaller towns, many of them clearly visible from the pyramid, the highest structure in the city. On the lake were canoes by the thousands, coming to and going from the city, crossing from one side to another by going under the frequent bridges on the causeways and along the canals that laced the city.

Cortés paid particular attention to the bridges. It would be easy for the natives to cut the bridges and leave the Spaniards trapped in the city, unable to retreat or to receive supplies.

Nor was this the only troubling thought. Just below them was another structure which had drawn their curiosity as they approached the main temple. It was a smaller pyramidal structure, on which were erected wooden racks holding human skulls, bleached white in the sun. One Spaniard calculated that the racks contained 136,000 skulls, all victims of sacrifice in the temple compound. The figure was probably far too high, but it was an awesome and thought-provoking sight.

Whatever uneasiness he may have felt, Cortés addressed his men confidently: "What do you think, gentlemen, of this great favor which God has granted us? After having given us so many victories over so many dangers, He has brought us to this place from which we can see so many big cities. Truly, my heart tells me that from here many kingdoms and dominions will be conquered, for here is the capital wherein the devil has his main seat; and once this city has been subdued and mastered the rest will be easy to conquer."

Cortés turned to Fray Olmedo and said he thought he would ask Moctezuma for permission to build a church here on the top of the main pyramid, with a cross that could be seen from throughout the city. As he had done on other occasions in response to Cortés's missionary impulses, Olmedo urged caution and patience.

So, instead of asking permission to build a Christian temple above the heathen temple, Cortés merely asked if he might not see the idols. After consulting with his priests — men whose garments and unkempt hair were stiff with dried human blood — Moctezuma agreed and led the Spaniards to the two oratories that stood at one side of the flat top of the pyramid. Each was about three stories high, the lower part of stone, the upper portions of wood. They were sacred to Huitzilopochtli, the god of war, and to Tezcatlipoca, the god of darkness and the infernal regions.

Bernal Díaz later recalled his impression of Huitzilopochtli: "It had a very broad face with monstrous, horrible eyes, and the whole body was covered with precious stones, gold and pearls. . . . The body was circled with great snakes made of gold and precious stones, and in one hand he held a bow, in the other some arrows. . . . Around the neck were silver Indian faces and things that we took to be the hearts . . . made of gold and decorated with many precious blue stones. There were braziers with copal incense and they were burning in them the hearts of three Indians they had sacrificed that day. All the walls and floor were black with crusted blood, and the whole place stank."

The other figure, he said, had "the face of a bear and glittering eyes made of mirrors. . . . It was decorated with precious stones, the same [as the other] . . . for they said the two were brothers. This Tezcatepuca [Tezcatlipoca] was the god of hell and had charge of the souls of the Mexicans. His body was girded with figures like little devils, with snakelike tails. The walls were so crusted with blood and the floor was so bathed in it that in the slaughterhouses of Castile there was no such stink. They had offered to this idol five hearts from the day's sacrifices."

The reverence with which Moctezuma regarded his idols brought home to Cortés and his men the gulf that existed between them. Moctezuma was just as awesomely devoted to his gods as the Spaniards were to their own. The creature comforts that they were enjoying, the geniality of their royal host, the gifts of gold and feathered finery were, by comparison, insignificant.

As the Spaniards left the temple compound to return to the palace of Axayacatl they saw something else that added to their uneasiness. Flanking each of the four gates to the compound were storerooms with vast quantities of weapons — bows, arrows, lances and the dreaded two-handed wooden swords with obsidian edges. It was easy to imagine that at a call from Moctezuma the city's tens of thousands of able-bodied men would drop whatever they were doing, destroy the bridges, rush to take up weapons and fall on the tiny army of intruders and kill them all — or take them alive to the temple and rip the hearts from their chests in front of the frightful idols.

TEN

Tenochtitlan (III)

CORTÉS WAS LATER to claim, in one of his dispatches, that he had planned from the outset to make Moctezuma his prisoner: "I decided to go and see him, wherever he might be . . . [and] I assured your Highness that I would take him alive in chains or make him subject to Your Majesty's Royal Crown."

But, as usual, his course was one of improvisation, seemingly determined as much by the sentiments of his followers as by Cortés himself. The first careful look at the city, and particularly the blood-stained temples, had left them all decidedly uneasy.

At first the Spaniards had been waited upon with great solicitude. Moctezuma kept up a constant flow of gifts. Although most of the food was strange, it was carefully and abundantly served to them. Bundles of grass and bouquets of flowers were brought to the horses.

But as the servants became increasingly familiar with the strangers and their obvious mortality they became unafraid, even surly, and the food diminished. The Spaniards' Tlaxcalan allies, as nervous in their traditional enemy's capital as were the white men, spread rumors that the Aztecs were planning to cut the bridges and spring the trap.

The men were somewhat distracted by the activity of build-ing a chapel within Axayacatl's palace. Cortés, frustrated in his desire to build a Christian shrine at the summit of the main pyramid, directed their efforts, and Moctezuma provided stonemasons to assist them.

In the process of building the chapel the Spaniards found in one of the stone walls what appeared to be a door that had recently been cemented over. The cement was chipped away. Beyond it lay a series of enclosed chambers. Cautiously the Spaniards entered and by yellow torchlight saw mounds of golden ornaments and jewels and stacks of gold bars. It was, by far, the richest treasure they had yet seen in this land of gold. On Cortés's orders the treasure was left in place, the door resealed. But the incident reinforced an idea that Cortés had been preaching from the outset: they were playing for very high stakes indeed — wealth such as few of them had even dreamed of. If the Aztecs turned on them it would be horrible enough to die. But to die leaving this untouched wealth would be particularly bitter. And while some of the men had been talking of retreat from the Mexican capital while retreat was still possible, to do so without taking this glorious treasure trove would be unthinkable, and removing it might provoke an assault by the natives.

Four of the captains and a dozen soldiers, including Bernal Díaz, came to Cortés to discuss their plight. "We said that we should seize Moctezuma if we valued our lives, and not wait another day," Bernal Díaz recalled. To which he said Cortés replied: "Do not think that I am asleep or that I do not have the same anxiety, but how can we seize such a great prince in his own palace surrounded by guards and warriors?"

Once more Cortés had managed to have his followers ask the question he wanted asked, and to which he had the answer already prepared. While in Cholula a few weeks earlier, he said, he had received word of an outbreak of hostilities near Veracruz. Juan de Escalante, left in charge of the rear base, had a message from Cuauhpopoca, the Aztec chieftain of

Nautla, which lay to the north of Veracruz, that he wished to come to the Spanish settlement and offer his allegiance to Spain. But in order to do so he needed an escort of Spanish soldiers to take him safely through the territory of the Totonacs who, with the Spaniards' encouragement, had already rebelled against the Aztec empire. Escalante had sent four soldiers to escort him. When they arrived in Nautla they were attacked. Two of them escaped and made their way back to Veracruz, but two were killed — and it was the heads of these that had been sent to Moctezuma for his personal inspection. Later it was learned that Escalante had mounted a punitive expedition against Nautla. The Spaniards waged a brief and successful war of reprisal against the Aztec outpost but took heavy losses: Escalante himself was one of the seven Spaniards fatally wounded. Dispatches sent to Cortés in Tenochtitlan by survivors indicated that Cuauhpopoca had been acting on Moctezuma's orders and added the alarming intelligence that the Totonacs, supposedly the Spaniards' allies, were showing signs of disloyalty.

The suggestion — or perhaps supposition — that Moctezuma was responsible gave Cortés the excuse for his most audacious act to date: the imprisonment of Moctezuma as a guarantee of the Spaniards' safety.

While the bolder of the conquistadors applauded the decision, everyone was apprehensive about the chances of such a daring stroke succeeding. Most of the men spent the night in the newly built chapel, saying prayers and confessing to Fray Olmedo — while Cortés paced back and forth in his apartment in Axayacatl's palace, planning his steps, move by move.

On the morning of their ninth day in the city Cortés and a picked group of captains and soldiers, all fully armed and armored, marched from their quarters to the palace of Moctezuma. They were, as before, greeted with courtesy, Again Moctezuma brought out gifts for his guests and he announced to Cortés that he wished to present to him one of his daughters. Cortés made his stock reply: that he already had the one wife allowed him by Spanish custom, and that anyway the girl had not been baptized as a Christian (these inhibitions later disappeared; the girl, christened Doña Ana, became part of Cortés's household for a time).

With the pleasantries out of the way Cortés got down to business and told the story of Cuauhpopoca and the allegation that Cuauhpopoca had been acting on Moctezuma's orders. Moctezuma denied it vigorously. Cortés insisted that Cuauhpopoca be brought to the capital for questioning. Moctezuma removed a jeweled amulet from his wrist and handed it to a courier, instructing him to use this symbol of authority to order Cuauhpopoca to come to the capital. But that was not all, Cortés said. Until Cuauhpopoca could be questioned and Moctezuma's responsibility determined, the Aztec ruler must come and stay with the Spaniards in their palace, not as a prisoner but as a guest. His presence there would guarantee that justice would be done; it would also be a convincing gesture of his goodwill toward the Spaniards and their king.

Moctezuma blanched at the thought of such an outrage. Then his face flushed with anger — and the guards Cortés had stationed at the doors fingered their weapons. Moctezuma protested the indignity of the suggestion. One of Cortés's captains, Juan Velázquez de León, angrily proposed that the Spaniards kill Moctezuma on the spot — an outburst which Marina dutifully translated for the beleaguered monarch. He nervously suggested an alternative: that one of his sons and two daughters be taken by the Spaniards as hostages. Cortés waved the suggestion aside.

Finally, with the strange fatalism that was rapidly destroying him as a ruler and as a man, Moctezuma agreed to go along. Cortés assured him that once the Cuauhpopoca matter was settled he could return to his own palace. With that Moctezuma, his royal litter surrounded by Spaniards, was carried out of his palace, across the temple compound and into his father's palace, which he had unwittingly permitted to become a Spanish fortress.

Moctezuma, in the Spaniards' custody, lived as he had before, with an army of courtiers, servants and entertainers, with sumptuous meals, luxurious baths, his wives and concubines. Apparently convinced that all of this was the will of his gods, he made the best of his embarrassing plight. He fraternized freely with the Spaniards. A Spanish page, Juan de Ortega, otherwise known as Orteguilla, was picking up the Nahuatl language and held conversations with the imprisoned king. Cortés learned the rules of a backgammonlike game called *patolli* Moctezuma liked to play and frequently gambled with him. Moctezuma seemed to be most pleased by losing, and when he won he distributed his winnings to the Spaniards who watched. He continued to make frequent gifts to his captors. When Bernal Díaz complained to Orteguilla that he had yet to receive any of Moctezuma's largesse, the page mentioned it to the king. Moctezuma thereupon presented Bernal Díaz with a gift of gold and cotton finery and a beautiful young girl, the daughter of one of his nobles. She was christened Doña Francisca and became Bernal Díaz's *naboría,* a word used loosely to describe either servant or bedmate. Another Spanish soldier made an improper demand of Moctezuma and behaved rudely. Cortés immediately condemned him to execution — and the wretch was saved only by the intervention of Moctezuma himself; he was flogged.

When Cuauhpopoca, accompanied by his son and fifteen noblemen, was carried into the capital in a litter ten days after Moctezuma's imprisonment, he was first questioned by Moctezuma. He then was turned over to Cortés, who questioned him some more. Cortés insisted that Cuauhpopoca had not only admitted killing the Spaniards but also claimed he was acting on Moctezuma's orders. He and his entourage were condemned to death by burning.

On Cortés's orders the weapons that had been seen in storerooms at the gates to the temple compound were gathered and piled high in the plaza. One Spaniard said there were fifteen cartloads of them — an unlikely estimate since there was not a cart, nor for that matter any kind of wheeled vehicle, in all of Mexico.

With the pyre built, Cortés appeared before Moctezuma and, with a stern expression, personally fastened iron shackles and chains to his arms and legs. It was a grave affront to the monarch. Until then he had told his courtiers and attendants that he was not a prisoner but instead only a guest of the foreigners, that he was living with them only because it amused him. Now there was no question of his status. Moctezuma wept and his attendants attempted to comfort him by thrusting their fingers inside the shackles to keep the iron from touching his royal flesh.

Cuauhpopoca, his son and fifteen accompanying nobles, bound hand and foot, were thrown on the pyre and the fire started. They suffered the agony without any sign or sound of pain. Natives thronged the plaza to watch the execution but gave no indication of grief or horror. Ceremonial executions had been commonplace in their lives, and the fact that the executioners were foreigners seemed to make no difference.

When the grisly business was finished Cortés returned to the chained Moctezuma, smiling and solicitous, and again with his own hands removed the shackles. He informed the unhappy man that he was now free to return to his own palace. Moctezuma declined, insisting that he preferred to stay where he was.

Whatever spirit Moctezuma had retained was now broken. He may have been afraid of the disrespect of his warlords, men intolerant of any human frailty, who would be particularly scornful of a king who had submitted meekly to such indignities. Or he may have simply realized that his old way of life was destroyed and beyond recall, presumably the will of his inscrutable gods.

Cortés and his men had achieved the greatest security they had thus far known in the Mexican capital. The execution of Cuauhpopoca and his entourage had demonstrated that the Spaniards were not to be trifled with; if any of the Indians

assumed to be friendly who had been left behind in the hot
country had been tempted to throw off their alliance with the
Spaniards they would now have second thoughts. Moctezuma
was completely in their hands and would issue whatever
orders the Spaniards dictated.

Cortés sent to the coast for hardware and rigging from the
scuttled ships, and with timber and skilled woodworkers provided by Moctezuma built three brigantines on the lake as
insurance of escape in case the Aztecs should cut the bridges.
Moctezuma was fascinated with the project.

Later he was taken on sailing excursions on the lake, often
visiting the island of Tepepolco, a royal hunting preserve, to
kill deer, rabbits and waterfowl.

Moctezuma also provided warriors to stand guard while
Spaniards prowled through his various residences, picking up
what they could in the way of movable valuables. With this
loot and the hoard in the treasure rooms of Axayacatl's palace,
the Spaniards now had a vast amount of wealth.

In addition, Cortés, on information supplied by Moctezuma, sent out scouting parties with native guides to look for
the sources of the gold. There were gold placers in streams and
rivers to the south, and before long the scouting parties came
back with impressive amounts of nuggets and gold dust. Other
gold came in tribute from communities in those parts of
Mexico dominated by the Aztecs. The amount of gold collected has never been precisely determined. Historian William
Hickling Prescott in 1843 calculated that it was worth $6.3
million; by late twentieth century prices for gold this would
mean at least twelve times that.

Whatever the amount it was the cause of much dissension,
both at the time and in later lawsuits brought against Cortés
by his former companions. There is general agreement, however, that after the royal fifth was set aside Cortés personally
claimed another fifth and still more to reimburse himself for
his expenses equipping the expedition and to repay Diego
Velázquez for the ships he had furnished. There were bonuses

for the captains, the priests and the men who had brought horses.

In the end the foot soldiers were said to have been allotted no more than one hundred pesos, a pitiful share of the vast treasure. Although Cortés had forbidden gambling, he vio-lated his own rule and so did most of the soldiers, playing with cards cut from drumskins and colored by hand. Most of the captains had native craftsmen make heavy gold chains for their personal adornment, and there were many quarrels over supposedly unequal shares.

Cortés did what he could to ease the unrest. He distributed gifts of gold from his own stock to those who were most vocal in their complaints — gold, the Spaniards said, is a great breaker of rocks. He adjudicated quarrels between his men. And he talked, always, of the greater treasure to come, once the country was entirely subdued. And to this task he was devoting his greatest attention.

After having had Moctezuma's artists draw maps of the coastline he determined that the most promising seaport for New Spain should be located at the mouth of the Coatzacolcos River, and he sent a detachment there under the command of Juan Velázquez de León to build such a port. An army of Indian workmen under Spanish command was dispatched to build a plantation for Charles V at Malinaltepec. Within a few months approximately ninety acres had been sown to maize, with other fields given over to beans and cacao.

But Cortés did not allow his gold hunting and his empire building — in all of which the captive Moctezuma was not only acquiescent but cooperative — to dull his alertness to danger. He had taken the head of an empire but he did not yet possess the empire itself. Moctezuma's regal authority, while still effective, had diminished. Finally Moctezuma him-self informed Cortés that his nephew Cacama, the king of Texcoco, just across the lake to the east, who had opposed permitting the entry of the Spaniards in the first place, was now talking rebellion.

There was good reason. The presence of the Spaniards

affronted everything Cacama believed in. Besides, he had been personally injured. The incorrigible Pedro de Alvarado had gone to Texcoco, demanding gold. When Cacama was either unwilling or unable to produce it, Alvarado had the young king seized and ordered his body scalded with hot tar — still without producing any gold. Cacama had become an even more bitter and outspoken enemy of the white invaders and was soliciting — and getting — support from other kings and lords.

Acting on Moctezuma's information, Cortés ordered Cacama to come to Tenochtitlan. Cacama defied the order, insulting both the Spaniards and his uncle, whom he described as an old hen. Moctezuma, on the instructions of Cortés, then sent a party of warriors to Texcoco. The young king was taken prisoner, brought to Tenochtitlan and put in chains.

Moctezuma's own brother, Cuitlahuac, king of Ixtapalapa, was similarly imprisoned for his part in the conspiracy. So were several other local rulers known to be or suspected of being conspirators. It was a damaging blow to the Tenochtitlan-Texcoco-Tacuba alliance, which had been the keystone of the Aztec federation.

At Cortés's urging Moctezuma summoned all other lords and rulers of cities and states subservient to Aztec rule. Not all of them came, but those who did were ushered into a reception hall in Axayacatl's palace. Here, with the proceedings being taken down by Pero Hernández, a Spanish notary, Moctezuma repeated the story he had already told the Spaniards, about his people being latecomers in this country and about the Spaniards being representatives of the great lord to whom it really belonged. He publicly declared his allegiance to this foreign rule and asked his vassals to do the same. Bernal Díaz, who witnessed the ceremony, said that Moctezuma's speech was marked with tears and sighs, and that the lesser lords wept as they listened to him and so, he said, did many of the Spaniards because "he was so dear to us."

At this point Cortés had achieved virtually total control.

With patient and prudent diplomacy and some treachery, he had won much. With a little more patience and prudence, his tenuous foothold might be quickly strengthened into a firm foundation of empire.

There was one problem — the incompatability of the Spaniards' Christian dogma and the harsh, demanding heathenism of the Mexicans. Moctezuma, characteristically, insisted that everything that had happened — the presence of the Spaniards, his own captivity, his fealty to an unknown foreign king — was all the will of the Aztecs' god, Huitzilopochtli. At the same time he indicated that whatever further the god might order would, in time, be conveyed to his people.

He had, meanwhile, resisted the best evangelical efforts of Cortés and the other Spaniards. He and his people were satisfied with their own religion and wanted none of the foreigners'.

Had Cortés been content to accept matters thus and trust to time, he might thereby have spared the lives of hundreds of his comrades and thousands upon thousands of natives. But this was not his way. He frequently violated holy writ in his personal life, but public display of his devotion to Christianity was another matter. It was rigorous and demanding. In one native town after another he tumbled idols from the native temples, raised crosses and erected figures of the Holy Mother — and would have done more of it had he not been restrained by Fray Olmedo. Olmedo was just as eager as Cortés, perhaps more so, for effective conversion of the natives to Christianity. But he was also more realistic, knowing that the change from one deeply-rooted belief to another was not a matter of the moment.

Aside from pagan idolatry, the most odious aspect of the Aztec religion from the Spanish viewpoint was human sacrifice and the attendant ceremonial cannibalism. Cortés had repeatedly ordered his prisoner Moctezuma to forbid further human sacrifice, but it had continued. In addition to the continued ritualistic bloodbaths, the hold of Huitzilopochtli and other idols on the will of Moctezuma and his people was a major obstacle to the conversion of Mexico not only to Christianity but to a manageable colony of Spain.

What happened next differs from one chronicle to another. Bernal Díaz, in most things a sharp-eyed observer, said only that Cortés asked the permission of Moctezuma to erect a cross and altar to the Holy Mother on the summit of the principal temple "and as time went on he would see how it would bring prosperity, health and good crops. With many sighs and an expression of great sorrow Moctezuma said that he would take it up with his priests, and after much discussion our altar . . . was put up apart from their damnable idols, and we all gave thanks to God."

Cortés, in one of his letters, made it more dramatic: "The most important of these idols, and the ones in whom they have the most faith, I had taken from their places and thrown down the steps; and I had those chapels where they were cleaned, for they were full of the blood of sacrifices; and I had images of Our Lady and of other saints put there, which caused Moctezuma and the other natives some sorrow . . . they believed that those idols gave them all their worldly goods, and . . . they would become angry and give them nothing . . . leaving the people to die of hunger."

Andrés de Tapia, one of the younger members of Cortés's company, left a more detailed account of the temple incident in his brief memoir of the conquest. Tapia said that on Cortés's orders he had climbed the main temple, followed by Cortés and eight or ten other Spaniards. With their swords they hacked their way through a hemp curtain, hung with golden bells, which shielded the entrance to the main oratory. Gazing at the idols within, Cortés exclaimed, "Oh God! Why do you permit such great honor to the Devil in this land? Look with favor, Lord, upon our service to you here." Through his interpreters Cortés lectured the attendant priests on the evil of their gods and the goodness of his. The priests laughed and said the Aztec people had risen in arms and were ready to die for their gods. Cortés sent for additional men and meanwhile gave way to an impulse. He seized an iron bar and, leaping high in the air, struck a smashing blow at the stone figure of Huitzilopochtli, tearing the gold mask away from the idol's face. By this time Moctezuma, under guard, had arrived at the temple and pleaded with Cortés not to injure the idols. The captive king suggested that the Spaniards might prepare a shrine to their deity at one side of the temple platform. Cortés refused this offer of separate accommodation and insisted that the Christian altar must occupy the place of Huitzilopochtli. Moctezuma once more retreated before the Spaniard's overpowering will and asked only that the priests be allowed to remove the idols peacefully. The stone figures, at least twice as large as men, were gently lowered down the sloping sides of the pyramid and the Spaniards set about scrubbing the blood-clotted temple. Finally they installed an altar with a figure of the Virgin and another of St. Christopher for, Tapia says, "we had no other images then."

Tapia added a footnote to the incident: "A few days later the Indians came bringing handfuls of corn that was green and very wilted, and they said: 'Since you have taken away the gods whom we asked for rain, now make yours give us rain or we shall lose our crops. . . .' Cortés assured them that it would soon rain, and asked all of us to pray for it. So the next day we went in procession to the tower, where a mass was held, and the sun was shining brightly. But by the time we left it was raining so hard that our feet were covered with water as we crossed the courtyard, and the Indians marveled greatly."

The rain-making episode tends to cast some doubt on all of Tapia's colorful details. Nevertheless, it could have happened. This was in April, 1520 — a time of year when rains occasionally do fall in central Mexico as a prelude to the rainy season.

But this was not the end of the incident. Moctezuma seemed to have been galvanized by the temple outrage into the sort of action he had been incapable of for the past five months. He summoned Cortés and addressed him with great gravity. The

Aztec gods were angered by the desecration and the Spaniards' greediness for gold, gold which belonged to the gods themselves. They would desert their people unless the strangers were expelled. Moctezuma's people were furious and were ready to rise and drive the Spaniards from their land. He urged Cortés to organize a hasty retreat with all his men and allies in order to save themselves. Otherwise every one of them would be killed.

But, while Moctezuma was being more forceful than he had been, there was still some hesitancy in his attitude toward Cortés and his army. In earlier times he had been a notable fighter, both in single combat and in command of field armies. He could easily have raised an army of one hundred thousand men in hours and could have crushed the invaders by the sheer force of numbers — without ever warning them. Instead he was displaying concern, even affection, for the men who had invaded and perhaps destroyed his empire.

Cortés said he and his army would leave if they were not wanted. But he needed time. They would have to build three new ships on the coast to take them away from Mexico and back to Spain. He would need many native workmen and craftsmen to fell trees and shape the timbers under the direction of his own shipwrights. The ships would be built as rapidly as possible, and when they were ready he hoped to take Moctezuma with him on the trip back to Spain to see the king.

The native workmen were provided and set off for the Gulf coast. Cortés was to suggest later that he had no intention of building the ships, and that he had his master carpenter, Martín López, build a scale model only as a false clue to his intentions.

The Spaniards cautiously kept the horses saddled and bridled at all times, and the men slept in their armor and their boots. Then came another summons from Moctezuma, who appeared happier than before. He unfolded before Cortés one of the native paintings that served as messages in his intelligence network. It had just come to him from the Gulf coast. It showed a fleet of eighteen Spanish ships anchored in a bay. Pictographic writing indicated the presence of eight hundred men, eighty horses and ten or twelve pieces of artillery. With what must have been some satisfaction Moctezuma pointed out that it would be unnecessary for Cortés to build new ships, and that they would have to delay no longer. He and his troops could return to Spain on the ships of their countrymen.

Many of the Spaniards shouted with joy and fired their muskets in the air in celebration of what they took to be good news. Surely this meant that reinforcements had come from Spain to help them strengthen their hold on this glorious but shaky dominion.

Cortés was overjoyed at first and exclaimed, "Blessed be the Redeemer for His mercies!" But he soon knew better. A courier arrived with a message from his rear base at Veracruz. The fleet was commanded by Pánfilo de Narváez, who had been commissioned by Diego Velázquez to seize both Cortés and the land he had conquered.

ELEVEN

Veracruz (II)

CORTÉS'S COURIERS, PUERTOCARRERO and Montejo, entrusted with the first big shipment of gold from Mexico, reached Spain in October, 1519, and dropped anchor at Sanlucar. Until then the loot brought back from the New World, while interesting and tantalizing, had been an insignificant return on the effort and expense that had gone into exploration and settlement: small amounts of rough gold, crude trinkets and exotic plants. But this shipment — gold bullion, the great gold and silver wheels, precious jewelry — was rich, beautiful and exciting. But there was trouble.

The confidential agent of Diego Velázquez in Spain circulated word that Puertocarrero and Montejo were mutineers against the authority of Governor Velázquez. As a result, the ship and its treasure cargo were seized and impounded in Seville by the Casa de Contratación, the administrative agency in charge of commerce with the Indies. Cortés's two envoys, however, were left at liberty, although virtually penniless. They went to Medellín, where they were joined by Martín Cortés, the conquistador's father. The three men then set off to find Charles V and report on Cortés, his accomplishments and the treasure.

The young king, surrounded by greedy Flemish courtiers, was preparing to go to Germany, where he had been chosen to succeed his grandfather, Maximilian I, on the throne. His principal concern at the moment was to raise enough funds in Spain to pay for a showy coronation in Germany. En route to the northern frontier of Spain he stopped at Tordesillas for a visit with his deranged mother, Joanna, and here, in March, 1520, Cortés's envoys caught up with him.

At about the same time the treasure had been released from impoundment in Seville and sent to the royal court. Charles, in gratitude for the badly needed gold, might have been disposed to grant Cortés the authority and autonomy the conquistador was seeking. But he was in a hurry to leave for Germany. And Bishop Fonseca, president of the Council of the Indies and a stout supporter of Velázquez, vigorously opposed it. As a result, no action was taken on the status of either Cortés or the great new province he had added to Spain's growing empire.

Meanwhile in Cuba, Velázquez, encouraged by his appointment as governor, was wasting no time in preparing reprisals for Cortés. At almost the same time that Puertocarrero and Montejo reached Spain, Velázquez had begun organizing an expedition to track down Cortés and bring him to justice. He at first announced that he would lead the expedition himself. He changed his mind, however. He said it was because the Indian population of Cuba was being swept by a smallpox epidemic and his responsibilities dictated that he stay at home and look out for his wards. Others found a more likely explanation in his obesity, which made any physical effort hazardous. As his surrogate he chose Pánfilo de Narváez, a tall, redheaded native of Valladolid, in whom Velázquez had confidence.

Narváez had come to Cuba from Jamaica in 1512 and, in command of a company of archers, had participated in Velázquez's conquest of that island, displaying the sort of ruthlessness toward the natives that Velázquez favored. Bartolomé de las Casas, who was in Cuba at the time, said that Narváez was a man of "good conversation and . . . habits, valiant in fighting Indians and should have been in fighting others," but, he added, "he was not very prudent. . . ." Others characterized him as indiscreet, arrogant, vain. However, his loyalty to Velázquez was unquestionable, and together the two men put together the largest armada that had yet been assembled in the Indies: eleven ships and seven brigantines.

Virtually every European settler in Cuba was either persuaded through the promise of rich rewards or forced into service, from hidalgos and landowners down to swineherds and mule drivers. Of seventy residents of the town of Trinidad, some fifty-odd were recruited, some of them unwillingly. Afterward it was said that at least one hundred fifty of the Narváez expedition were eager to get to Mexico, not to serve Narváez but to join Cortés and share the fortune he was accumulating for his men.

In the end Velázquez and Narváez had nine hundred men, including substantial numbers of crossbowmen and harquebusiers, eighty horses and perhaps a dozen artillery pieces. Narváez and his fleet sailed from Guanaguanico, at the western tip of Cuba, in March, 1520 — just about the time Puertocarrero and Montejo were pleading Cortés's case at Tordesillas.

Notice of the expedition had, meanwhile, come to the attention of the Spanish *audiencia* in Española, a body responsible for judicial and civil administration in all the Indies. The report caused alarm. Cuba would be left almost devoid of Spanish settlers. Worse, Narváez seemed prepared to launch a civil war against Cortés, a war which would destroy whatever illusions the Mexican natives might have about Spanish unity, but which might also jeopardize Spain's dominion in this new, rich country. Accordingly, the *audiencia* sent one of its members, Lucas Vázquez de Ayllón, to Cuba to try to halt the expedition.

Ayllón arrived in Guanaguanico and ordered Velázquez

and Narváez to halt the venture. He was ignored. He then followed the fleet in his own ship, hoping to prevent any outright warfare. Not long after arrival in Mexico, Narváez had Ayllón arrested and put aboard his own ship and expelled to Española, where a report on Velázquez's insubordination was prepared and sent to Spain.

Narváez had landed on April 23 at San Juan de Ulúa, as had Cortés before him. And like Cortés before him, he immediately set about establishing a town, which he proposed to call San Salvador, and a town government. While this was going on he was visited by three of Cortés's men, who had unsuccessfully been scouting the country for gold mines, on Cortés's orders. The three were beguiled by Spanish food and wine and informed Narváez that Cortés's followers were disappointed in their fortunes and disillusioned with Cortés's promises to share the considerable wealth he had amassed. They also told of the tiny Spanish garrison — made up mostly of the ill and elderly — at Veracruz, only a little distance to the north.

Gonzalo de Sandoval had recently taken over the command at Veracruz. Sandoval was one of the youngsters of the Cortés expedition — he was then barely into his twenties — but he was a valiant fighter and a steady and responsible man. When he received word of the arrival of Narváez's fleet he sent those of his men who were weak and ill to the mountains, while he remained behind with those most fit for combat.

Narváez sent a priest named Guevara, a notary and four others of his men to Veracruz to demand submission. When they entered the town, it seemed deserted. They noted the gallows, stopped in the church to pray and then found the most imposing residence, which they correctly presumed would be that of the garrison commander. Entering and meeting a cool Sandoval, Father Guevara ordered the notary to read the orders which Narváez had dictated. Sandoval insisted that they first offer proof of authority from the Spanish crown. When the notary tried to proceed without doing so, Sandoval had the six men seized. He summoned Indian carriers — he had maintained excellent relations with the natives — who bundled the six men into net hammocks, swung them on their backs and set off for Mexico, where they were to be delivered to Cortés.

Cortés meanwhile had been faced with a series of complex problems and hard decisions. Moctezuma had been in communication with Narváez and had sent him gifts of gold and cotton cloth. His envoys had been assured by Narváez that the latter's mission was to capture and punish Cortés and to restore Moctezuma to freedom.

Cortés was in a precarious position. He had been warned that if he did not leave Tenochtitlan the Aztecs would take up arms against him and drive him out. The forces led by Narváez were superior in numbers and arms. If Cortés waited for Narváez to attack him in Tenochtitlan, the Aztecs would probably rise at the same time. If, on the other hand, he marched against Narváez, what should be done with Moctezuma? If he took the royal hostage with him, his leverage for controlling the capital would be gone. Left behind, Moctezuma might rally his people.

While Cortés pondered these alternatives word was received that Father Guevara and the other five Spaniards arrested by Sandoval were approaching. Carried in their net hammocks on the backs of the surefooted native porters, they had covered the several hundred miles from Veracruz in four days. To save them the indignity of being carried into the capital like trussed-up pigs, Cortés sent an escort out to meet them, had them mounted on horses and brought into the city with ceremony. He personally greeted them with warmth and kindness, apologized for the rough treatment they had suffered at the hands of his subordinate, showed them some of the great stores of gold and other treasures and gave them samples. He learned from them that Narváez's command was shaky, many of his men restless. Cortés sent a conciliatory letter to Narváez in Guevara's care, urging that they unite in the conquest of this rich land. And he warned that a show of disunity among brother Spaniards would only invite the destruction of all of them. Liberally supplied with gold, Guevara and his companions made their way back to the coast. Narváez was outraged by Cortés's message. But the import was not lost on his men. This Cortés had much gold and seemed to be generous with it.

Cortés soon had another similar message under way, this one carried by Fray Olmedo, who had become a valuable counselor in temporal as well as spiritual matters. Olmedo was to treat with Narváez diplomatically. He was also to talk with Andrés de Duero, once secretary to Diego Velázquez, who was with Narváez. And perhaps more important than the messages Olmedo was to deliver were the quantities of Aztec gold he was to distribute where it would do the most good. Olmedo's message angered Narváez just as had Guevara's — and one of his aides loudly declared that he personally would cut off Cortés's ears and have them broiled for his breakfast. Olmedo also found Duero, gave him personal messages from Cortés, and mingled freely with other officers and men in the Narváez company, liberally distributing gifts of gold and leaving the unmistakable impression that there was much more where this had come from.

Sandoval, meanwhile, not only kept his little garrison at Veracruz on the alert, but also sent some of his men, disguised as Indian fruit peddlers, to spy on the Narváez camp; two of them managed to steal a pair of Spanish horses on one of their intelligence forays. Some of Narváez's men deserted and joined forces with Sandoval.

Narváez had discovered, as Cortés had before him, that the barren sand dunes of San Juan de Ulúa were an unsuitable campsite and had moved on to Cempoala, where he and his men occupied the principal temples. They helped themselves freely to what the Indians had, and some of his officers appropriated Indian girls who had been presented to Cortés and his men but who had been left behind in their parents' custody.

Sandoval kept up a steady stream of reports to Cortés. With

these and his own sure instincts Cortés developed a battle plan which, on the face of it, seemed to invite disaster. If it succeeded it would be the most daring maneuver yet. He would march against Narváez with a skeleton force, leaving behind in Mexico a garrison to guard Moctezuma. In command of it he placed Pedro de Alvarado, a man whose recklessness was a problem but who was both feared and respected by the natives. He would leave with Alvarado one hundred forty men, approximately two-thirds of the force available. Some of those being left behind were unfit for marching and combat duty, and many of them had pro-Velázquez leanings. He would take with him only men of tested loyalty. "By great gifts of gold and promises to make us rich he persuaded all of us to stand by him," wrote Bernal Díaz. "We were happy with . . . the gold Cortés gave us, as though he had taken it from his own property and not from what should have been our share anyway."

Moctezuma offered to send five thousand of his warriors with Cortés, but the offer was politely declined. Such a force of "allies" could annihilate his own little force in minutes. Setting forth from Tenochtitlan, Cortés had only seventy men and five horses.

But Cortés had sent orders to Juan Velázquez de León, who was scouting the Coatzacoalcos region, and Rodrigo de Rangel, who was doing the same thing in the area of Chinantla, to join him at Cholula. Both men did, and their troops increased Cortés's force about threefold. The loyalty of Velázquez de León was significant. He was a kinsman of Diego Velázquez. Narváez had sent him an invitation to join the invading force, which he might have been expected to do, but he had stoutly declined. And Rangel brought word that the people of Chinantla hated the Aztecs and were eager to help Cortés in any way they could. They were good warriors, wielding twelve-foot lances with great efficiency. Cortés sent back word to Chinantla asking them to send him two thousand warriors and also to manufacture three hundred more of the long lances, tipped with copper instead of the usual stone or volcanic glass. He reasoned that the lances would be useful in the hands of his own foot soldiers in meeting Narváez's cavalry charges.

He also asked his old friends, the Tlaxcalans, to provide him with a native fighting force of five thousand men. The Tlaxcalans, who had been eager to fight with him against the Aztecs, had no desire to join him in fighting other Spaniards. Instead they sent their best wishes and ten loads of fowl.

Later Cortés was joined by Sandoval and as many men as he could spare from the garrison at Veracruz. With this addition Cortés's strike force had grown from 70 men to 266, if one counted both Canillas the drummer and Beger the fifer. There were a few muskets and crossbows. Few of the men still possessed metal armor such as their enemy would be wearing. It had been discarded one place and another. Most of them now wore the native-style quilted cotton, both as body protectors and helmets. Most of it was ragged and dirty — but what Cortés's men lacked in military smartness they made up for in gold ornaments. Gold chains bounced on their chests and encircled their arms.

They marched east as fast as they could. En route they met Fray Olmedo, returning from his mission to Narváez's camp. He conveyed to Cortés his impressions of the lack of discipline and loyalty among Narváez's men. A man of good humor with a gift for mimicry, he acted out for the troops the pompous and arrogant mannerisms of Narváez and his principal officers. Cortés laughed with his men. Nonmilitary things such as foolish pride, stupidity and avarice often decided military destinies as much as did cold steel, gunpowder, spirited horses and raw courage.

In the enemy camp, Cortés's men later learned, Narváez was also having some thoughts on military destinies. "You think Cortés could be so brave," he sneered to his men, "to come to this camp with the three cats he commands?" Still he was thinking of ways to avoid an armed clash.

On their march across the bleak tableland before descending to the tropics, the tiny army was intercepted by Andrés de Duero, Cortés's old friend, who had come out to meet him. He brought a message from Narváez. If Cortés would submit to his authority, he would provide him ships in which to leave this country. Cortés replied that if Narváez would show him a royal commission authorizing his acts he, Cortés, would submit. But inasmuch as Narváez represented only Diego Velázquez, Cortés and his men would obey only orders from the king they served. He composed a reply summoning Narváez and his men to appear before him and acknowledge allegiance to the king whom Cortés represented — a representation which the king had only recently heard about from Puertocarrero and Montejo in faraway Tordesillas.

In private conversation with Duero, Cortés promised to share with him the gains of the conquest. Duero willingly provided even more detailed information than Olmedo had about conditions, strengths and weaknesses in Narváez's command. Cortés thanked him but warned that when the attack came he would expect continued cooperation; if not, Cortés would himself run his old friend through with a sword.

On the march from the tableland down into the hot country of the coast, Cortés and his men were overtaken by a party of Indian porters from Chinantla with the three hundred extra-long copper-tipped lances that had been ordered. The Spaniards held a short drill with the long weapons, seeing how they could be best handled to halt or unseat a charging horseman. Then they marched on to the river that flowed just to the south of Cempoala. They were by now within a few miles of the enemy. They learned later that Narváez, with some cavalry and foot soldiers, had ridden out on the other side of the river looking for them, but had given up and returned to camp. The river was high and swift from the late spring rains when Cortés and his men reached it. There were periodic downpours, and the river would continue to rise. Darkness came, and the men stretched out on the soggy ground to rest.

They had killed some deer and wild pigs during the day's long march, but cooking fires would have betrayed their presence. The night sky was alternately bright with moonlight and dark with rain clouds. Cortés, without raising his voice above a normal speaking tone, delivered another one of his speeches combining emotion, patriotism, piety and the promise of alluring rewards. He reviewed the dangers and hardships they had been through together, and the threat to them that the Narváez venture posed. Then, as Bernal Díaz remembered it, he "began to flatter us and our courage in battle, saying that before we were fighting to save our lives, and that now we had to fight for both life and honor, for now they were coming to capture us, drive us from our houses and rob us of our property . . . and he said that in war prudence and intelligence were more important in conquering the enemy than mere daring."

Andrés de Tapia recalled a somewhat different thing. "I am but one man," he quoted Cortés as saying, "and can speak for no more than one man. Offers have been made to me that were advantageous only to myself; and because they were disadvantageous to you I have not accepted them. You have heard what they say; therefore the matter rests with you. Whether it is your inclination to fight or seek peace, speak your mind and no one shall hinder you from doing as you wish. These messengers of ours have told me in confidence how it is being said in the enemy camp that you are deceiving me in order to place me in their hands. . . . Say what is on your minds." Tapia said the men shouted their loyalty and urged him to go on. Cortés continued: "There is a saying in Castile . . . 'may the donkey be killed or else the one who goads him.' That is my opinion. For I see that anything else we might do [other than attack] would be a disgrace to us all."

Whereupon, Tapia said, "we gave a shout of joy, saying 'Hurray for our captain and his good judgment!' Then we picked him up and carried him on our shoulders until he had to beg us to let him go."

Cortés gave precise instructions. This company was to go

direct for Narváez, who had his quarters at the top of the principal temple. That company was to overcome the artillery that was positioned to defend the temple. All was to be done quietly and by stealth.

Holding their weapons above their heads the attack force forded the swift river, struggling against the strong current. Two of the men were swept away and never seen again. As quietly as they could they moved through the jungle, the damp night air heavy with tropical perfume. At a short distance from Cempoala they captured two sentries. Questioned, one of them admitted that Narváez knew they were in the immediate area and was planning to sweep it at dawn. He also corroborated details of defense arrangement, location of quarters, artillery placement and other things that Cortés already knew. The other sentry, however, escaped and ran into the village shouting alarm. Cortés ordered his fifer and drummer to sound a charge.

It was astonishingly easy. Although the alarm had been raised, Narváez's men were confused and disorganized. The winking of thousands of fireflies was mistaken for musketeers' matchlocks and they feared they were being attacked by an inexplicably large force. Narváez's artillerymen had plugged the matchholes of their cannon with wax to keep the powder dry. Before they could get them unplugged, Cortés's men were upon them and turned the weapons on their masters. A party led by Sandoval dashed up the steps of the main pyramidal temple where Narváez had his quarters. A torch was thrown onto the thatch roof of the oratory-turned-command-post. In the ensuing fight a pike struck Narváez in the face and put out an eye. According to Bernal Díaz, who was in the attacking party, the unhappy commander shouted, "Holy Mary, help me! They have killed me and put out my eye." Some of the Spaniards, hearing it, shouted, "Victory, Victory! . . . Victory for Cortés, for Narváez is dead."

Narváez was not dead, but there was little fighting spirit left in either him or his men. When Cortés approached him,

Narváez said, "Congratulations, my lord Cortés, for capturing me." Cortés, matching arrogance with arrogance, replied, "The least deed I have done in this land is to take you."

Narváez lost seventeen men in the fighting, Cortés only two. Narváez and his captains were taken in chains to Veracruz. The rest of the army willingly, even eagerly, came over to Cortés's side. Weary and without having slept, Cortés sat on a bench in the morning, an orange gown draped over his armor, and received the homage, hand kisses and pledges of loyalty from Narváez's men. He was magnanimous with them. He ordered his own men to return to them horses and weapons they had seized. Bernal Díaz, who was forever envious of those of his companions who owned horses, lamented that he had to surrender a horse, two swords, three poniards and a dagger. To those of his men who grumbled about such orders, Cortés explained that they were overwhelmingly outnumbered by the men they had conquered and that it was well to cultivate their favor.

As a further security measure he divided up his vastly increased forces. One hundred twenty men, of whom twenty were Cortés veterans, were sent on a settling expedition to the Coatzacoalcos region under the command of Diego de Ordaz. A similar group was sent under Juan Velázquez de León to the Pánuco region to the north. Two hundred more were detailed to San Juan de Ulúa to strip Narváez's fleet of sails, rigging and hardware. He also sent dispatches to Alvarado, telling of his victory.

The two thousand warriors from Chinantla marched into Cempoala on the day after the brief battle, three abreast with pike carriers on the outside, bowmen in the center. Cortés thanked them, rewarded them with beads and sent them home again. Cortés's mastery over these obviously fierce warriors impressed his new Spanish allies.

Then came an alarming message from Alvarado. The Mexican people had risen in arms and were besieging the Spaniards.

It took some time for it to become clear what had happened. Even before Cortés had left the capital, the Aztecs had, through Moctezuma, requested permission to celebrate their annual Toxcatl festival in honor of Huitzilopochtli, the war god. Cortés was assured there would be none of the usual sacrifices, only dancing and singing. The festival would continue in the great temple plaza for ten days, from May 9 to 19. It began five days after Cortés left for the coast. Warriors and nobles, dressed in their greatest finery, thronged into the plaza, singing and dancing to the ominous throb of their huge wooden drums. Traditionally two young men — destined for this end since the year before — were prepared for sacrifice. For a year they had lived in luxury, and for the past few weeks they would have received the tender attentions of a group of the city's prettiest young women. They would climb the steps of the temple, breaking the flutes on which they had learned to play, be stretched on the sacrificial stone and have their hearts cut out by the priests. Afterward their heads would be impaled on poles in the plaza while the music and dancing went on. Precisely what happened on this occasion was unclear.

One version was that Alvarado's men seized the two victims-to-be, pressed confessions from them and then attacked the crowd in the plaza to prevent retaliation. Another was that when the festival and religious fervor were at a climax, Alvarado and his men, fully armed, went into the plaza and at a signal fell on the celebrants with swords and pikes, slaughtering hundreds, perhaps thousands — like a replay of Cholula. One of Alvarado's men later quoted his captain as saying, "We have fallen on those knaves — as they meant to fall on us we struck first."

Whatever the cause and sequence of events, the results were quick and terrifying. The Spaniards were driven back into their palace by an army of outraged Aztecs. They were attacked with stones, arrows and spears. Torches were set to inflammable parts of the palace. Alvarado freed Moctezuma from the chains and, with a dagger pointed at his throat, compelled the Indian ruler to speak to his people from the roof, trying to calm them. The mob shouted insults at their king and continued to shoot arrows and throw stones.

The attack went on for a week. Breaches had been made in the palace wall. Part of it had been burned. The brigantines on the lake were set on fire. And supplies of food and water were shut off. Seven Spaniards had been killed, many more were wounded and all were starving. Then, after a week, the violence ceased. The attackers had had word of Cortés's victory over Narváez and knew that he would be marching on Mexico with a greatly increased army. Still, an army of a thousand or so Spaniards could be easily overcome by the hundreds of thousands of men the Aztecs could field. The white men would be permitted to reenter the city and would be trapped. So the attacks were halted while the Indians rested and prepared for the ultimate battle that would end their shameful domination by these strangers.

Cortés reacted quickly, decisively. He recalled the troops that had been sent out under Ordaz and Velázquez de León. The march inland was difficult. Most of his new army was unaccustomed to the extremes of temperature, the difficulty of the mountain ranges. In Tlaxcala finally Cortés was relieved to find his old allies still friendly and faithful. Food and other supplies were provided in abundance. So was a large troop of Tlaxcalan warriors who were eager to battle the Aztecs. Cortés marched on in high spirits.

TWELVE

Tenochtitlan (IV)

WHEN CORTÉS APPROACHED Tenochtitlan on Sunday, June 24, 1520 — St. John's Day — his manner was more lordly than before. He had won a critical victory over his own countrymen at Cempoala against staggering odds. Now he was at the head of a column of 8,000 Tlaxcalan warriors, more than a thousand Spanish foot soldiers and almost a hundred mounted horsemen — by several times the largest force he had yet commanded. Some of the Spaniards were his own veterans, but the greater part was made up of men who had arrived with Narváez and had, with varying degrees of eagerness, come over to Cortés's side.

During the tiring march from the humid, scented air of the tropics to the thin, dry air of the valley of Mexico, Cortés had beguiled his new followers with lavish tales of Aztec treasures, the hospitality and generosity of the natives, the reverence with which they treated the Spanish "god," the beauty and luxuriousness of a city which he dominated. Alvarado, whom he had left in charge of the captive city and the captive king, may have had some trouble with the natives. But it would all be straightened out when he, Cortés, returned. The natives would be out to greet them with flowers and gifts.

At one of the bridges on the Tepeyac causeway a horse ridden by one of his men slipped between two beams, broke a hoof and had to be destroyed. Blas Botello, one of Cortés's veterans, who had long been regarded by his companions as an astrologer, magician and soothsayer, said that it was a bad omen, that it was a sure sign of greater mishaps to come.

Cortés shrugged off the prediction, but his high spirits were sagging for other reasons. As the marching column drew nearer the city there were no flowers, no brilliantly costumed crowds eager for sight of the Spaniards. In fact there were no people at all unless one counted a corpse seen hanging from a beam of a burned-out house. The sound of marching men and the clatter of hooves echoed between the empty buildings eerily. Piled in one of the streets was a heap of tortillas and live fowl, tied together, apparently left as a gift of food. As they neared the center of the city there were more burned buildings. A gun was fired in the air as a signal, and an answering gunshot was heard from the palace of Axayacatl. The column hurried on to the battle-scarred palace.

Cortés, usually congenial and easy-mannered, had become short-tempered and irritable. Moctezuma, informed of his arrival, requested that Cortés come to see him. Cortés refused to see "that dog of a Mexican." Later he turned down Moctezuma's request for permission to call on him, despite the fact that Moctezuma, as a peace gesture, offered to have a statue of Cortés cast in solid gold. It was a decision he later was to regret; he would never again see Moctezuma as he had known him before. He was equally testy with Alvarado, his close friend and valued captain. Alvarado, through his rapacity and impetuousness, had repeatedly caused trouble for Cortés since the very beginning of the expedition, but had always been forgiven. But the Toxcatl massacre had graver consequences than any of Alvarado's earlier misdeeds, and Cortés questioned him closely. He learned of the damages resulting from the Aztecs' retaliatory anger. Many of the causeway bridges were out. The three brigantines that Cortés had built and launched on the lake had been destroyed. Supplies of food and water for the band of Spaniards in Axayacatl's palace had been cut off and the markets were closed. But there had been no attacks on the Spaniards for the past fortnight; instead there was only an ominous quiet in the deserted streets.

On Monday morning Cortés sent word to Moctezuma that he must order the markets reopened and normal commerce resumed. Moctezuma replied that he was powerless to do such a thing so long as he was a prisoner, and if he could not be released — of which he was certain — then release of one of the other royal captives might achieve the desired end. And he recommended his brother Cuitlahuac, king of Ixtapalapa. Cortés consented. It was one of a series of errors that Cortés, usually foresighted, was to commit in that fateful last week of June. Cuitlahuac, vigorous and young — he was then about thirty-five — had opposed the Spaniards' entry into Mexico from the beginning, and Spanish chains had heightened his militancy. Furthermore he was a legitimate claimant to the throne of his helpless brother, since succession was more often from brother to brother than from father to son.

Cortés also sent off a dispatch rider with messages for his men in Veracruz and his allies in Tlaxcala, informing them of his safe arrival and of the situation prevailing in the capital. Within half an hour the messenger had returned, bruised by stones and thoroughly frightened. Approaching the palace through the once-deserted streets was a huge army of Aztec warriors, fully armed and keyed to a fever pitch, brandishing weapons and shouting.

The gates to the palace compound were secured and artillery positions manned. Within moments the vast army struck the palace like a dark wave and the air was filled with stones, arrows and spears. Flaming arrows were shot into the thatch that served as roofing for most of the buildings. Inside, the Spaniards blazed away with firearms. Those who were not manning battle posts were trying to put out the flames. Because water was scarce, dirt was thrown on the fires. In order to

prevent the spread of the conflagration, part of one of the protecting walls had to be dismantled, and the danger to the besieged men was thereby aggravated. It was, by far, the fiercest
fighting the Spaniards had yet experienced in the New World.
The Aztecs showed no fear of the foreigners' firearms. Although the cannon could cut wide swaths in the masses of
warriors, the gap would immediately be filled by equal numbers, just as fierce, just as furious. From a distance they used
slings to hurl stones with deadly accuracy. Closer they used the
atlatl, or spear thrower. Closer still they used the terrible two-
handed wooden swords and three-pronged darts with cords
attached for their retrieval. They screamed and howled con-
tinuously and shouted threats at their enemies — they would
be killed on the sacrificial stones, their arms and legs eaten,
the rest thrown to the animals in Moctezuma's zoo, or, alternately, the flesh of the Spaniards was not fit to eat; their entire
bodies would be thrown to the animals, which had been
starved for days to insure good appetite. The Tlaxcalan allies,
who were thin, would be placed in cages to be fattened before
being taken to the sacrificial stone.

Several times the Spaniards rushed out of the gates to see if
they could break the encirclement. They took heavy losses in
both men and horses and were thoroughly awed by the feroc-
ity of a people. Bernal Díaz, who in his memory often made
mistakes in names, dates and details, reconstructed the horror
of these days with convincing fidelity: "These combats lasted
all day long, and even during the night so many squadrons
fell on us, hurling javelins, stones and arrows in volleys, as
well as odd stones, that with those who fell during the day and
those that fell in all the patios [of the palace] it was like
wheat on the threshing floor. . . . [Another day] our captain
decided to sally out at dawn with all of us and Narváez's men,
and take our cannon and muskets, and endeavor to defeat
them or at least make them feel our strength and valor better
than we had done the day before. I tell you that when we
made this decision the Mexicans had decided to do the very

same. We fought very well, but they were so strong and had so many squadrons that relieved each other from time to time that even if 10,000 Trojan Hectors and as many more Rolands had been there they would not have been able to break through. . . . Cannon and muskets were not enough against them, or hand-to-hand fighting, or killing 30 to 40 of them every time we charged, for they still fought on in as close formation and with more energy than at the beginning. If at times we gained a little ground or part of a street, it was because they made as though to retreat so that we would follow and then they could fall on us and cut us off. . . . They had wooden drawbridges between their houses; when we were going to cross to burn their houses they raised the bridges so that the only way we could cross was through deep water. And we could not endure the rocks, stones and javelins they hurled from the rooftops. . . . Some three or four soldiers . . . who had served in Italy swore to God many times that they had not seen such furious fighting in any they had encountered between Christians, against the artillery of the King of France, or even the Grand Turk."

The constant rain of rocks, even boulders, from the flat rooftops was a grave peril. Helmets and skulls were crushed. Weapons were shattered and made useless.

After several days of fighting Cortés ordered the construction of some war machines which his men called *tortugas* (turtles). Each was like a huge box, roofed over, mounted on rollers and with enough portholes for the twenty-five men inside to discharge their weapons. With these, it was thought, the Spaniards could cope with the rock throwers of the rooftops. Cavalry charges were to clear the way for the *tortugas*, but the ungainly vehicles, although built of heavy timbers, were of little use against the fury of the attackers. They were soon wrecked, and the accompanying cavalry suffered heavy casualties from spear thrusts.

A large group of the attack force had assumed a position at the top of the principal temple, virtually overlooking the palace in which the Spaniards were barricaded. Cortés first ordered one of his men to take an assault party up the pyramid and overcome the warriors at the summit. The first assault failed and Cortés decided to direct the attack himself. His left hand had been wounded — the resulting paralysis of two fingers was permanent — but he tied his shield to his left arm, took his sword in his right and led a charge across the pavement of the temple courtyard, slippery with blood, and up the face of the pyramid.

To climb the pyramid unimpeded was difficult enough. The stair steps were twice as high as they were wide. At each terrace the climbers had to circle the huge structure to reach the next flight of stairs. The distance covered, then, amounted to nearly a mile, and at every step the Spaniards advanced under a steady rain of missiles from above, while a rear detail guarded the base of the pyramid to avoid harassment of the attackers. Finally Cortés and his men reached the summit and were immediately engulfed in hand-to-hand fighting. Spaniard and Aztec would grapple and, often, tumble over the edge, their bodies bounding down the steep sides to be crushed on the pavement below. Gradually, however, the white man's steel overcame the Indian's flint and glass, and the platform at the top was thick with bodies and blood.

In the fighting thus far there had been a number of "miracles" of the sort that marked the adventures of sixteenth century Spaniards everywhere. Desperate for water, the Spaniards had dug in the courtyard of Axayacatl's palace. And although they were surrounded by the brackish salt water of Lake Texcoco, they uncovered a spring of sweet water, which they interpreted as a sign of divine providence. St. Mary and St. James were said to have appeared in the fierce battle, St. James wielding a sword and St. Mary throwing dust in the eyes of the Indians. A cannon aimed at the Indian attackers did not respond to the fire of the matchlock. It was abandoned while the cannoneers engaged in hand-to-hand fighting. When the mass of attacking Indians became still denser, the cannon fired of its own accord, taking a great toll. Similarly, it was said that the Indians had attempted to remove the image of the Virgin Mary from the shrine the Spaniards had installed in an oratory at the top of the pyramid. The image, the Spaniards said, could not be budged. Furthermore, those Indians who had laid hands on the holy image bore permanent stains on their hands. How the Spaniards could know either of these things was not explained.

Several times Cortés attempted to hold truce parleys with the attackers. To his argument that all the slaughter and suffering was unnecessary, that victory was impossible, the Aztecs retorted that they would go on fighting, even if twenty-five thousand of their warriors had to die for every Spaniard killed.

Cortés was still having nothing to do with Moctezuma. But Fray Olmedo and Cristóbal de Olid called on the unhappy monarch and urged him to intercede for peace. Reluctantly Moctezuma agreed. He donned his royal finery and mounted one of the parapets on the wall around the palace. His people were massed below him. The cries and whistles and the throb of war drums ceased when he raised his arms. The Aztecs, who had never dared look at their ruler directly, now gazed at him openly and listened to his words. Moctezuma said that he was not a prisoner, that he had been willing to come and stay with the Spaniards in the interest of public good. If the Aztecs would put down their arms, the Spaniards would leave Tenochtitlan if allowed to do so peacefully.

At first there was awed silence, then murmuring and then shouting. Moctezuma was denounced as a weakling who had abandoned his people to become a wife of the Spaniards. A bow was drawn and first one arrow and then many sped through the air toward Moctezuma, who was inadequately protected by the Spaniards' shields. Arrows were followed by stones hurled at great velocity from slings — and the battle was on again.

Legend — as pervasive on the Aztec side as on the Span-

ish — had it that the first arrow was shot by a nephew of
Moctezuma, Cuauhtemoc, or Falling Eagle. True or not,
Cuauhtemoc was a man with whom the Spaniards would have
to reckon. He was fearless and determined and was destined to
be the last ruler of the Aztec empire, and possibly the most
heroic.

Neither Cuauhtemoc's arrow nor those of others did any
harm, but the stones did. One of them struck Moctezuma on
the temple, and the Spaniards carried his limp body back
inside the palace.

Moctezuma, with no will to live, lingered for three days,
refusing food and medical attention. When the bandages were
placed on his wounds he tore them off. Similarly, he rejected
Fray Olmedo's efforts at conversion and baptism. Then he
died. As Bernal Díaz remembered it, "Cortés . . . and all of
our captains and soldiers . . . wept for him. There were men
among us who wept as though he had been our father . . .
considering how good he was." Cortés was much briefer in
recalling the event in a dispatch to the king: ". . . He died
three days later. I told two of the Indians who were captive to
carry him out on their shoulders to the people. What they did
with him I do not know; only the war did not stop because of
it, but grew more fierce and pitiless each day."

Spanish accounts, on the one hand, and Mexican, on the
other, differ not only on the disposition of the body but also
on the cause of death. Most Mexican chronicles agree that
while Moctezuma was felled by a stone his death was caused
either by strangulation or by daggers wielded by the Spaniards. One account added the ugly detail that the final and
fatal wound was a sword thrust up his rectum.

Cortés knew that if the Spaniards remained where they
were they would either starve or be killed. Rations were down
to a handful of maize per day per man. Accordingly, Cortés
began reconnoitering the shortest of the causeways, the one
leading west to Tacuba, a distance of slightly more than two
miles. There were eight gaps in the causeway from which the

bridges had been removed. But men on horseback could jump
some of them and either wade or swim the others. Alvarado,
Sandoval and a few others went clear to the end, to Tacuba
itself. They came riding back carrying bouquets of flowers —
to be chided by Cortés for their frivolity.

Meanwhile the Mexicans had launched an attack on the men
on the causeway, and Cortés himself fought a rearguard
action, protecting the retreat of his men toward the palace.

On his return he was visited by Mexican emissaries, who
informed him that if he would only release their high priest,
who was being held prisoner, they would discuss truce terms.
Cortés freed the captive priest. He had no sooner done so than
the attack was resumed with greater fury than before. It
became apparent that the Mexicans needed the priest to con-
secrate the succession of Cuitlahuac to Moctezuma's throne,
and they had no intention of ending the war, then or ever.

Cortés went ahead with plans for an evacuation. A portable
bridge was built that could be thrown across a break in the
causeway and, once the troops had passed, could be picked up
by Tlaxcalan porters and carried to the next gap. He gave
marching orders. Two hundred foot soldiers were to go in the
vanguard under the command of Sandoval, and supported by
about two dozen horsemen. The rear guard was given to the
command of Juan Velázquez de León, whom Cortés trusted,
and Pedro de Alvarado, whom he liked but on whom he could
not always rely. Cortés himself took command of the center,
composed of the rest of the troops and cavalry, the freight,
including treasure, most of the artillery, various captive Aztec
lords who might be useful in future negotiations, and a son
and two daughters of Moctezuma whom that unfortunate man
had entrusted to Cortés's care. One of them, who had been
baptized and christened Doña Ana and thereby made eligible
for a Christian gentleman's bed, had shared Cortés's apartment and was said to be pregnant by him.

One of the many problems was the accumulated treasure.
Many individual soldiers already wore chains and collars

wrought of Aztec gold, but there was much more: the king's fifth, Cortés's own considerable share, and the shares owned by various captains, and much more, melted down into bullion and stored in a specially guarded room of the palace. Cortés delivered the king's fifth, or at least part of it, to royal officers whom he had appointed, and gave them a sturdy horse of his own to carry it and a guard of trusted soldiers to protect them. As for the rest — a vast quantity — he invited the soldiers to help themselves, but cautioned them against trying to carry too much, since what lay ahead would demand both strength and agility. Few heeded him. The Narváez men in particular were avid in loading themselves down with heavy gold bars. Cortés veterans were more cautious. Bernal Díaz recalled that he took only several of the jadeite pieces so highly valued by the Indians, reasoning that they would, in the uncertain future, be more useful for bartering than would gold. And they were easily carried.

Blas Botello, the astrologer who had interpreted the crippling of a horse as an evil omen for the Spaniards and whom Cortés had ignored, now made another prediction. The Spaniards, he said, would safely escape only if they left the city in the middle of the night. This time Cortés paid attention, and departure was scheduled for midnight Saturday, June 30.

The rainy season had begun, and while the rain usually ceased in the early evening, on this night it continued as a heavy drizzle accompanied by thick fog. Promptly at midnight the gates of the palace were eased open and the Spaniards crept out stealthily, struggling to keep their footing in the streets made slick by blood and rain. Four hundred Tlaxcalan porters carried the portable bridge and went ahead to lay it across the first gap in the causeway. The vanguard moved out, crossed the bridge, and the central division, commanded by Cortés, prepared to cross. But before they could do so the deep-throated war drums began to sound, conch shell trumpets blared and, in a matter of seconds, the Indian population was attacking as fiercely as ever, perhaps more so. On both sides of the causeway were canoes. From them warriors shot arrows and hurled spears. Some Indians abandoned the canoes and climbed up the causeway to grapple with the evacuees. Many of the Spaniards, encumbered with their gold, fell into the water, were seized, pulled aboard the canoes and quickly gutted with native knives. Part of the rear guard — estimates of their number ranged from a few score to several hundred — concluded that it was hopeless and retreated to the palace, where they hoped they would be secure. There they were soon overcome by the Indians and quickly taken to the top of the pyramid where, by torchlight, they were sacrificed in full view of their struggling companions on the causeway.

Meanwhile the portable bridge had become wedged in one of the gaps under the weight of so much traffic and had to be abandoned. But the gaps were slowly filling up with human and horse flesh, and the survivors picked their way across by walking on the bodies.

One of Cortés's men, Francisco de Aguilar — who late in life became a Dominican friar — described the scene that occurred after the loss of the portable bridge: "There was no way to cross the remaining canals — which were a good 12 feet wide and full of water — except that our Lord provided us with the Indian men and women who carried our baggage. As these entered the first canal they drowned, and the heap made a bridge for those on horseback to pass over. In this way we kept pushing the loaded bearers in front of us, reaching the other side over the bodies of the drowned, until we had crossed the rest of the canals. And in the confusion of drowning Indians some Spaniards also were lost."

In the rear guard, Juan Velázquez de León was dead. Alvarado, his sorrel mare dead from a spear thrust, was said to have used his lance to vault across one of the unclosed gaps. Alvarado later denied it, but the "leap of Alvarado" became firmly fixed in the legends of the conquest and even became a place name in the city that was to be built later.

Cortés rode back and forth, as far forward as Tacuba and

back as far as the last stragglers, cheering, scolding, ordering, encouraging, hacking at the thickly massed Indians with his sword. At dawn he dismounted and sat under a tree at Popotla, a village near Tacuba, surveyed the appalling human wreckage that lay behind him, and wept. In just the one week since he had reentered the city he had, for the first time, been beaten, humiliated and cruelly punished.

Cortés was later to report that he had lost from one hundred fifty to two hundred men during the *noche triste,* the sad night. One of his companions estimated the Spaniards' loss at eleven hundred and fifty. Bernal Díaz made a probably more realistic estimate of eight hundred fifty Spaniards dead (and a subsequent muster at Tacuba showed four hundred twenty-five survivors out of a total of twelve hundred and fifty).

The Tlaxcalan allies, both warriors and porters, took far heavier losses; at least four thousand of them and possibly as many as eight thousand died.

All but twenty-three of the Spaniards' horses were dead (a horse was reckoned to be worth three hundred foot soldiers), and almost all the rest were wounded. The artillery was lost. So was most of the treasure, including the king's fifth. And so were some of Cortés's most valued companions: Velázquez de León, Francisco de Morla, Francisco de Salcedo, his favorite page, Juan de Salazar, and one man who was known only as "Lares the fine horseman." Ironically, Botello, the astrologer and soothsayer who had specified this particular time as advantageous for a retreat, was one of the victims. So were children of Moctezuma who had been in Cortés's care — the pregnant Doña Ana was one of them — and various noble prisoners including Cacama, the rebellious young king of Texcoco.

The remnants of the army proceeded into Tacuba and took refuge in a temple that provided a view of the surrounding plain. Here the men rested and dressed their wounds as well as they could. They remained until midnight and once more took up the march, headed for Tlaxcala, which the Spaniards hoped would have remained loyal and friendly. Periodically the ragged column was harassed by bands of Indians. Many of the walking wounded lagged behind and were killed — or were simply left behind to die of their wounds and exhaustion. Some of the stronger ones saved themselves by holding onto the tails of the few surviving horses, most of which carried two riders. Cortés maintained discipline as well as he could. When one starving foot soldier cut the liver from a dead Indian and ate it, he was hanged on the spot.

Skirmishes continued, and in one of them Cortés suffered another wound. Still, there were surprisingly few Indians, and when the Spaniards reached the plain of Otumba they knew the reason: they had all congregated there. It was a vast and colorful army of hundreds of thousands in full battle array, waiting for the white men they knew must pass here en route to Tlaxcala.

For the Spaniards the odds for survival were dismal — even if they had been rested, well fed, well armed and healthy. Cortés called a halt and ordered his men to eat whatever food they still had in order to gain strength. He then formed his small force into a square with the wounded and ill on the inside, horsemen on the outside, and waited for the shock of the attack. It soon came.

Francisco de Aguilar described the scene: "Then, although he was in tears, he drew himself up and exhorted and encouraged us like a brave captain. . . . Artillery and harquebuses we had none, for all of it had been lost, but the Lord mercifully abated His wrath and looked upon us with favor, because as Cortés battled his way among the Indians, performing marvels in singling out and killing their captains . . . distinguishable by their gold shields, and disregarding the common warriors, he was able to reach their captain general and kill him with a thrust of his lance. . . . [Cortés] fell twice to the ground and found himself back on his horse again without even being aware of who had helped him up. The other Spanish captains, on horseback, also performed marvelous feats of valor in order to escape the death that stared them in the face. . . . We foot soldiers under Diego de Ordaz were completely surrounded by Indians, who almost had their hands on us, but when . . . Cortés killed their captain general they began to retreat and gave way to us. . . ."

Other sources indicated that the Indian leader, variously identified as Cihuatl and Ciuacoatl, was actually struck down by the lance of Juan de Salamanca, who then handed the fallen chief's standard to Cortés. But all the men agreed on Cortés's bravery and leadership. López de Gómara said: "Never had there been a more notable feat of arms in the Indies . . . and all the Spaniards who that day saw Hernán Cortés in action swear that never did a man fight as he did, or lead his troops, and that he alone in his own person saved them all."

The vast native army faded away and the Spaniards, wearier than ever, resumed their march across the arid plain toward Tlaxcala. They found a spring where they could drink, bathe and dress their wounds.

Finally they reached Tlaxcala. The Tlaxcalans grieved over the number of their own people who had been lost in Tenochtitlan; of the various noble young women who had been given to Spanish captains only the one who belonged to Pedro de Alvarado had survived. The heavy loss of warriors and porters had touched almost every family in the little nation. The Tlaxcalans chided Cortés for his failure to take them seriously when they had warned him of the duplicity of the Mexicans. But they nevertheless received the Spaniards warmly and took them into their homes to recover. Cortés was suffering from a fever, but after a chip of bone was removed from his scalp he began to recover and consider the future.

THIRTEEN

Tenochtitlan (V)

BARELY RECOVERED FROM his wounds, Cortés wrote to his king that he had decided that this new country should be called New Spain of the Ocean Sea, and that he had so christened it in the name of the king. And he added: "I intend . . . to return to that country and its great city, and . . . that it will shortly be restored to the state in which I held it before, and thus all our past losses shall be made good."

These were brave words for a man who had just survived the lowest point in a mercurial career. The loss of men, horses and weapons had been catastrophic. Those who survived were almost all wounded, some fatally, and some would be crippled for life. Most of them begged Cortés to return to Veracruz, where they could either receive reinforcements or withdraw from the country. The Aztecs had proven that the Spaniards were vulnerable and mortal. Word of this would spread to other provinces; the disillusioned natives could easily block all trails and passes between Tlaxcala and the coast. So ran the arguments for retreat.

"I told them," Cortés wrote, "that I would not abandon this land, for, apart from being shameful to myself and dangerous for all, it would be great treason to Your Majesty; rather I resolved to fall on our enemies wherever I could and oppose them in every possible way."

In less than three weeks after he had led his ragged little remnant of an army into Tlaxcala, Cortés led them out again on a campaign of aggression. The target was the rich agricultural province of Tepeaca, thirty-odd miles southeast of Tlaxcala. Tepeaca was an Aztec ally. Also, it straddled the best route from Veracruz to the interior of Mexico, the route by which reinforcements would come — if they ever came. Somewhat earlier, a party of a dozen Spaniards, en route inland to join Cortés, had been killed by the people of Tepeaca. This provided Cortés with an excuse.

The Spaniards alone were an insignificant force, but they were followed by several thousand Tlaxcalans, as eager as ever for war and spoils.

The campaign was quick and brutal. Within a few weeks forty towns had been sacked and burned, and estimates of the number of Tepeacans killed ran as high as sixty thousand — the greatest slaughter of the conquest to date. Surviving males were given to the Tlaxcalan allies to either enslave or kill. Women and children were enslaved by the Spaniards. They were branded on cheeks and lips with the letter G (for *guerra*, war) and portioned out, one-fifth for the king, one-fifth for Cortés, the remainder to be sold. In addition, Cortés ordered his men to turn in whatever native servants they had accumulated in the Tepeaca campaign. Bernal Díaz complained that when this was done the young and pretty female servants invariably became the property of Cortés and his captains, while only "the old ones and the ruins" were returned to the troops.

Cortés felt called upon to justify his imposition of slavery. To the king he wrote: "I made certain of them slaves of which I gave a fifth part to Your Majesty's officers, for, in addition to having killed the aforementioned Spaniards and rebelled against Your Highness's service, they are all cannibals. . . . I was also moved to take those slaves so as to strike some fear into the people . . . there are so many people that if I did not impose a great and cruel punishment they would never be reformed."

The use of cannibalism as justification was curious. The Spaniards had encountered it virtually everywhere they had gone. It was practiced by their friends, the Tlaxcalans, and other Indian allies. One account of the Tepeaca campaign, probably exaggerated, said that at one time the Tlaxcalans had fifty thousand cooking pots over their campfire, preparing meals of their victims' arms and legs. Another noted that while the Spaniards did not practice cannibalism, they often fed the bodies of battlefield victims to the small dogs the Mexicans raised for food. And the Spaniards had no objection to eating the dogs.

The Spaniards' campsite in Tepeaca was converted into a town, Segura de la Frontera, to guard the Veracruz road and to serve as a rear base for the operations that lay ahead. From here Cortés sent Martín López, a shipwright and carpenter, to Tlaxcala with orders to build a fleet of thirteen brigantines for use in the assault of Tenochtitlan. In the mountains of Tlaxcala were forests of pine and oak. Tlaxcalan woodcutters and craftsmen would be the labor force. Canvas for sails, rope for rigging, anchors, chains and other hardware salvaged from the Spanish ships on the coast were ordered brought up from Veracruz.

Spanish reinforcements began to arrive. One party of thirteen, sent out by Diego Velázquez to find out what had happened to Narváez, was tricked ashore at Veracruz by the port captain and imprisoned. The port captain then burned their ship and persuaded the little group to join Cortés in his Tepeaca campaign. A week later the same tactics were used on a second party which, while smaller in number, was well supplied with arms and ammunition.

Other ships began to arrive, sent out by Francisco de Garay, governor of Jamaica, to support a short-lived settlement on the

Pánuco River to the north of Veracruz. One ship brought one hundred fifty men, another fifty, and all were persuaded to join Cortés. The master of a merchant ship from Spain, heavily laden with food and military supplies, diverted his vessel from Cuba to Veracruz, correctly anticipating a livelier market. The port captain bought the entire cargo, and many of the seamen, lured by gaudy tales of gold, deserted and marched inland.

Cortés was once more at the head of something resembling an army rather than a party of Spanish refugees followed by a horde of native scavengers.

Cortés also had a secret ally: smallpox. The disease had been brought to Mexico by one of Narváez's men. Most of the Spaniards were already pockmarked and immune. But to the native population the disease was devastating. The Spaniards insisted that the natives' addiction to frequent bathing lowered their resistance and hastened the spread of the epidemic. Whatever the reason, the Indians died by the thousands.

Tenochtitlan was hard hit. The city stank with the bodies that no one had the strength to bury or burn. The most notable victim was Cuitlahuac, who had directed the last furious attack on the Spaniards and then, later, succeeded his brother Moctezuma. He died of smallpox after only eighty days on the imperial throne. He was succeeded by Cuauhtemoc.

Shortly after Cuauhtemoc's accession to the throne, Cortés moved the bulk of his army from Segura de la Frontera back to Tlaxcala. He had directed numerous forays into surrounding territory and now controlled the area from the peak of Orizaba on the east to Cholula on the west. In Tlaxcala he checked on progress in the construction of the fleet of brigantines. He wore mourning for the death of his friend Maxixca, another smallpox victim. He oversaw the "knighting" of Maxixca's thirteen-year-old son, a Christian ceremony save for the anointing of the youth with the heart of a just-sacrificed Aztec spy. Meanwhile still more reinforcements and supplies were coming up from the coast.

On December 26 Cortés held a review of his forces and published a list of ordinances to govern them: no gambling, blasphemy, brawling, quarrels over quarters, insubordination, attacking without orders or abandoning assigned positions. Also, all booty must be turned in to Cortés or his designated representative.

Two days later Cortés launched a preliminary phase of the campaign of reconquest. His army was divided into four squadrons of ten horsemen each, and five hundred fifty foot soldiers divided into nine companies, plus nine pieces of artillery.

On New Year's Eve they were in Texcoco, the Athens of the Mexican world, and settled in the palace of Nezahualcoyotl, the poet king. Texcocan royalty and nobility had fled. Cortés appointed a new ruler and set about converting the city — just across the lake from Tenochtitlan — into an advance base. Some Texcocans were added to his army of native allies. Others were put to work digging a canal to facilitate the launching of the brigantines. Cortés was soon off again, marching on Ixtapalapa, Cuitlahuac's city, the key to the southern approaches to Tenochtitlan.

There was bitter fighting at Ixtapalapa. But the native forces were finally overcome and the Spaniards occupied the city — almost to be trapped. The natives cut the dike that protected the city from the waters of the lake, and the Spaniards had to flee in disorder to escape drowning.

The Spaniards went as far as Cuauhnahuac (later Cuernavaca), a semitropical town beyond the mountains that lay south of Tenochtitlan. The town was protected by deep ravines which the Spanish foot soldiers could cross only by climbing one overhanging tree and leaping to one on the other side — a difficult feat when they were loaded with arms and armor.

Having subdued Cuauhnahuac, the Spaniards returned to

the great valley of Mexico and fought a prolonged and bloody battle with the people of the lacustrine town of Xochimilco, a place of lush and colorful gardens. Here it was noted that Spanish swords, captured by the Aztecs on the *noche triste,* were being used as points on native lances and were effective against the white invaders and their horses. At one point Cortés was thrown from his mount and was about to be captured when he was rescued by a strong-armed Tlaxcalan Indian. The identity of the Indian was never established, and Cortés was later to credit his rescue to the intercession of his patron, St. Peter.

The fighting at Xochimilco lasted for several days and if it proved anything it was that reconquest was not to be easy. The Spaniards burned part of the town and secured some loot — which did little more than encumber their movements. Some Spaniards were captured by the natives and hauled off in triumph to Tenochtitlan, where they were sacrificed in a great celebration. Cuauhtemoc sent the severed heads, arms and legs of the Spaniards to outlying cities and states as further grisly proof that the Spaniards were not invincible.

Cortés and his men withdrew to the deserted town of Coyoacan and then to Tacuba, the city at the end of the westward causeway from Tenochtitlan. It was here that the retreating Spaniards had regrouped after their flight.

By now Cortés had seen all approaches to the city and had tested the temper of the defending Aztecs. He had suffered heavy losses and would suffer some more. Cortés and some of his men climbed to the top of the highest temple in Tacuba and looked out across the water to the prize city, Tenochtitlan. Its pyramids, towers, palaces and flat-roofed houses glistened in the sunlight, only a few miles away. Canoes and barges loaded with food swarmed around the island city. Warriors in their finest regalia thronged the causeway, daring the Spaniards to come out and fight. The mood of the conquerer was pensive, even sad. He realized that this beautiful city, which he had hoped to seize intact as the crowning achievement of his adventure, probably could not survive the demands of victory.

Cortés returned to Texcoco. There was much to cheer him. Still more reinforcements had arrived. And the hulls of the thirteen brigantines, carried piecemeal on eight thousand human backs from the mountains of Tlaxcala, a distance of almost fifty miles, were being reassembled and rigged. The launching canal, twelve feet wide and twelve feet deep and more than a mile long, was ready for them. It had been completed in fifty days by a work force of forty thousand Texcocans.

But there was also upsetting news. Through an informer he learned of a plot to assassinate him. While he was at his dining table he would be handed a letter which purportedly had just arrived from his father in Spain. While Cortés was opening it — which he could be expected to do with great eagerness and concentration — the conspirators would fall on him and stab him to death. This plot originated among Narváez's men — as had so much of the past dissension. Cortés seized the chief conspirator, Antonio de Villafaña, a friend of the Cuban governor, Diego Velázquez, and promptly had him hanged from a window. He also obtained a list of the fellow conspirators. After memorizing the names, he destroyed the list and announced that if there had been a list Villafaña must have swallowed it before his execution. This materially cooled anti-Cortés sentiment in the ranks, at least for the time being.

In his frequent talks to his augmented army Cortés laid stress on both piety and practical matters: "The principal reason for our coming to these parts is to glorify and preach the faith of Jesus Christ, even though at the same time it brings honor and profit, which infrequently come in the same package." He also stressed that past injuries the Spaniards had suffered at the hands of the Aztecs must be avenged and the various abominations eradicated — human sacrifice, idolatry, cannibalism and sodomy. Among his listeners the likelihood of personal gain registered most strongly. One of his men said that he led them "to believe that each one of us would be a count or duke and one of the titled. With this he transformed us from lambs to lions, and we went out . . . without fear or hesitation."

By mid-May, 1521, all was in readiness for the final assault. Cortés, who until now had been little more than a guerrilla leader, a soldier who had profited mainly through happenstance, now played the role of a field marshal, weighing tactics and strategy, disposing his forces with careful calculation. He had a considerable force at hand. In addition to the casual — and in some cases unwilling — additions to his forces, three ships had come from the island of Española, in response to his messages, bringing more than two hundred men, more than seventy horses, and a large supply of arms and ammunition. A ship from Spain had brought more men, a royal treasurer, Julian de Alderete, and a Dominican friar, Pedro Melgarejo de Urrea, who had a supply of papal bulls, enough to clear the Spanish adventurers of any atrocities they might commit in the name of Christianity. Altogether Cortés had more than nine hundred Spaniards, roughly one-tenth of them mounted, more than a hundred of them armed with crossbows or harquebuses, three heavy iron cannon and fifteen smaller bronze artillery pieces, with shot and cannonballs and more than a thousand pounds of powder. It was at least twice the size of his original assault force, better armed and equipped. And for every Spaniard there were several hundred native allies, many of them armed with copper-tipped lances of Spanish design but native manufacture. There were at least fifty thousand Tlaxcalans, and there were many other thousands from Texcoco, Chalco, Tepeaca, Cholula and Huejotzingo. López de Gómara estimated the total fighting force of Spaniards and Indians at two hundred thousand — which, while no one knew for sure, was probably about the same size as the Aztec force defending Tenochtitlan. In addition, the Spaniards were supported by fifty thousand or so Indian laborers, who would carry supplies, build shelters, roads and bridges.

Pedro de Alvarado was given thirty horsemen, one hundred seventy foot soldiers, three artillery pieces and thirty thousand Tlaxcalan Indians and assigned to Tacuba to seal off the westward approach to the city — and also to destroy the aqueduct that brought spring water into Tenochtitlan from the hill of Chapultepec. Olid was given approximately the same number and assigned to Coyoacan to guard the southern approaches. Sandoval was given a somewhat smaller number and ordered to complete the destruction of Ixtapalapa, which had remained troublesome, then to join Olid at Coyoacan. Cortés assumed command of the fleet of twelve brigantines (one of the original thirteen was found to be defective). Each carried a crew of twenty-three men, one artillery piece and six harquebuses or crossbows, and could be propelled by either sail or oars.

Cortés sailed with his little fleet about June 1. The crewmen could see, around the shores of the lake, signal fires spreading the word that the attack was under way.

At the rocky island of Tepepolco, where Cortés had once taken Moctezuma on royal hunts, the heights were defended by Aztec warriors, shouting defiance. Cortés anchored his ships, led his men in storming ashore and up the heights, where they wiped out the little garrison. While this was going on, a fleet of a thousand or more war canoes was seen approaching from Tenochtitlan. Cortés and his men boarded the ships and, with oars ready, waited for action. The fleet of canoes drew up, and the two forces stared at each other. Suddenly an east wind sprang up, filling the brigantines' slack sails, and the Spanish ships plowed into the mass of frail canoes, shattering them like matchwood, while the Spaniards used lances, crossbows and muskets on those who survived drowning.

Cortés sailed on to the point at which the main causeway from Ixtapalapa to Tenochtitlan was joined by the causeway from Coyoacan. Here Cortés established his advance command post.

Alvarado and Olid had reached Tacuba. And although the

aqueduct which supplied Tenochtitlan with water from Chapultepec was heavily guarded they seized and destroyed it. Olid then continued his march along the western shore of the lake to Coyoacan.

Sandoval found the final destruction of Ixtapalapa an easy matter, quickly done. Cortés had planned for Sandoval to join forces with Olid at Coyoacan. Now he changed his orders, sending him instead to take the northern causeway leading from Tenochtitlan to Tepeyac.

With this the Aztec capital was surrounded with no access or egress except by water. The Spanish brigantines were expected to control canoe traffic — the only way in which the city could be supplied with food and fresh water.

The Aztecs were soon bringing in canoeloads of supplies by night to escape detection. Furthermore they invented tactics to neutralize the Spaniards' naval superiority. Canoes loaded with well-armed warriors would be concealed in rushes that bordered the lake. Nearby, stout stakes would be driven into the lake bottom. Other canoes would then decoy the brigantines into shallow water, where they would become hung up on the stakes, with neither oars nor sails effective. Then they would be attacked fiercely. But the brigantines continued to be powerful weapons. When the Aztec troops were thickly congregated on one of the causeways a brigantine would come alongside; its deck-mounted artillery could then rake the causeway with murderous effect.

With all the Spanish and allied forces in their assigned places, a general assault was launched on Tenochtitlan on June 9. The southern, or Ixtapalapa, causeway and the western, or Tacuba, causeway were in Spanish hands, and Sandoval was poised on the northern, or Tepeyac, causeway to shut off retreat. The bridges over gaps in the causeway had been broken by the Aztecs and barricades erected on the city side. But with native labor troops to fill in the gaps and with supporting fire from the brigantines, the Spaniards advanced steadily until they were in the heart of Tenochtitlan, the great plaza and the temple compound. The temple itself was attacked and taken and the idols rolled down the sides of the pyramid. But while the Spaniards were thus occupied, the Aztecs regrouped, brought up reinforcements and counterattacked fiercely in the plaza. The Spaniards retreated in some confusion, the native allies often impeding their withdrawal.

On the following day the Spaniards made another assault and again penetrated to the heart of the city. Here the imperial aviary, Moctezuma's pride and one of the handsomest buildings in a city of palaces, was sacked and destroyed by fire.

By the next day all the causeway gaps had been opened again and barricades erected by the city's defenders. The Aztecs discovered that the Spaniards made no effort to guard the causeways during the night. They would mount night attacks to distract attention and move in labor corps to reopen the causeway breaks and rebuild the barricades.

Certain of the Aztec warriors became familiar sights. Cuauhtemoc was active in all sectors, giving signals with his conch shell trumpet. A giant Aztec warrior named Tzilacatzin was dreaded by the Spaniards. His only weapon was his sling; with it he could hurl stones with deadly accuracy and effectiveness. Because of his huge size he was an easy target — but the Spanish crossbow arrows seemed to have no effect on him.

Cortés craftily arranged for various of his Indian followers to approach Cuauhtemoc and announce that they had had a change of heart and now wanted to fight on the Aztec side. Cuauhtemoc agreed, putting the turncoats in the rear ranks. Once the battle was under way the Spaniards' allies would turn their weapons on the Aztecs, attacking from the rear.

Despite such treachery and the obvious superiority of Spanish weapons, the defenders were punishing the invaders cruelly. The Aztecs were more eager to take captives than they were to kill. On one day eighteen Spaniards were captured. On another a fierce assault, led by a warrior named Tlapanecatl, ended in the capture of fifty-three Spaniards, four horses and hundreds of the Spaniards' Indian allies. On still another Pedro de Alvarado was ambushed and saw five of his men taken off as captives. The postscripts to these captures were particularly harrowing for the Spaniards. The great war drum would begin to throb from the main temple. Then the captives, stripped to their white skins and with their heads wreathed in flowers, would be led up the steps of the pyramid, one by one. Moments later their bodies, minus the hearts, which had been ripped out and offered to Huitzilopochtli, would be tumbled down the pyramid's steps.

Bernal Díaz recalled it with a shudder: "I considered myself a good soldier and had the reputation of being one. . . . [But] when I saw my companions sacrificed, their hearts taken out still beating, their arms and legs cut off, I was truly afraid that one day it might happen to me. They had already seized me twice to take me to be sacrificed, and it pleased God that I escaped . . . but ever since then I feared death more than ever. Before going into battle a kind of terror gripped me. . . ."

On June 30 a massive all-points assault was launched in the hope of seizing the market district of Tlatelolco. Cortés urged his captains to advance only after filling gaps in the causeway behind them. But Julian de Alderete, the royal treasurer who had only recently joined forces with Cortés and had been given a command, disregarded the order. He found the Aztec forces melting away in front of him and charged ahead at full speed. When he was far extended the Aztecs attacked furiously. Retreat was almost impossible, and other Spanish units had to go to Alderete's rescue.

In the bloody fighting that followed, Cortés was pulled from his horse and would have been carried away for sacrifice if he had not been rescued by Cristóbal de Olea. Olea cut off the arm of one of the captors with a single blow of his sword and killed four more before being killed himself. Cortés was wounded in the leg.

The attacking forces of Alvarado and Sandoval were also

driven back, Sandoval suffering three wounds. Seventy Spaniards were taken captive and, while Cortés upbraided Alderete
for disobeying orders, the war drum thundered from the
temple, signaling another mass sacrifice. It was the first anniversary of the *noche triste*.

The Spaniards' plight was growing serious. Many of the
Indian allies had been deserting, the warriors, the laborers
and porters and even the women who prepared the Spaniards'
sparse meals — mainly tortillas, cactus fruit and a thin sauce
made of chiles. Soon only the Tlaxcalans remained with the
Spaniards. Worse, supplies of gunpowder and shot were running low. So were arrows for the crossbows. On many nights
Cortés himself joined in the work of filling causeway gaps and
making new arrows for the crossbows, fitting points and fasten-
ing feathers to the shafts with a native glue known as *zacotle*.

An old soldier from Seville persuaded Cortés that the answer to the stalemate was a catapult, which he would build if
Cortés supplied materials and labor. It would be able to hurl
stones great distances, wreck buildings and kill hundreds of
defenders. Cortés agreed, and the huge structure took shape,
an assemblage of beams, wheels and coiled ropes. The Indian
allies shouted to the Aztecs that this secret weapon would be
their final undoing. Finally it was finished, loaded with a big
stone, cranked up and released. The stone flew straight up in
the air and dropped back, wrecking the machine. The Aztecs,
who had been watching apprehensively from a safe distance,
laughed and shouted insults.

Meanwhile Cortés had come to the conclusion that the city
must be utterly razed, that nothing could be left standing to
shelter the defenders. Temples and palaces had already been
burned. Now block by block and house by house, wooden
beams were put to torch and walls of stone and adobe pulled
down, the native allies doing the work while the Spaniards
fought off the Aztecs. While the work of destruction went on
the Aztecs taunted the Indians doing the work: "Go ahead!
No matter how it ends you will be forced to rebuild the city,

either for us or for them." What had been a beautiful island
city became a flattened wasteland of rubble, seven-eighths of it
destroyed. Only the Tlatelolco district remained more or less
intact.

The Spaniards' fortunes began to change. A ship put in at
Veracruz with a full cargo of needed arms and ammunition.
And the native allies who had deserted sensed that the Aztecs
were doomed and began to return to the Spaniards' encamp-
ments. Cortés had been stern in the past, inflicting death for
desertion (the young Xicotencatl had deserted Alvarado's
forces and had been executed), but now he welcomed them
back with courtesy and promises of good things to come.

Survivors were crowded into the Tlatelolco section. They
were reduced to eating tree bark, roots, bitter grasses gathered
from the lakeshore, bits of leather and deerskin, lizards, corn-
cobs and dirt. Although the Aztecs were accustomed to eating
human flesh after ceremonial sacrifices, they did not eat the
corpses of those fallen in battle, and the streets were piled
high with corpses. Cortés led one predawn sortie to observe the
enemy scavenging for food and killed, by his own account,
eight hundred of them, "all unarmed, principally women and
children."

On the night of August 11 there was a downpour of rain, and
at the height of the storm a ball of fire was seen in the sky,
moving rapidly to the south, trailing sparks. Then, suddenly,
it plunged into the dark waters of the lake and disappeared.
The besieged Indians viewed it as a portent of certain doom.

Two days later a brigantine commanded by Garci Holguín
was patrolling the lake near Tlatelolco. He saw a large canoe
setting out from shore. Its passengers were finely dressed. He
came alongside and ordered the canoe to halt. The obvious
leader stood and with sad formality announced that he was
Cuauhtemoc, demanded that he be taken to Cortés. Holguín
took the captive king aboard and sped to the headquarters of
the conqueror.

Cuauhtemoc behaved decorously and unselfishly. He asked

first for the safety of his wife and companions. Cortés described the encounter: "As I had no desire to treat Cuauhtemoc harshly, I asked him to be seated, whereupon he came up to me and, speaking in his language, said that he had done all he was bound to do to defend his own person and his people, so that now they were reduced to this sad state, and I might do with him as I pleased. Then he placed his hand upon a dagger of mine and asked me to kill him with it; but I reassured him saying that he need fear nothing. Thus, with this lord a prisoner, it pleased God that the war should cease, and the day it ended was Tuesday, the feast of Saint Hippolytus, the thirteenth of August, in the year 1521."

The capture of Cuauhtemoc brought an end to the natives' resistance. A deathly silence spread over what had been a great city, and had recently been a tumultuous battlefield. The stench of the thousands of unburied, unburned corpses sickened the victors. It was estimated that of three hundred thousand Aztecs in the city at the outset of the siege, only sixty thousand had survived. Most had fallen in battle, but perhaps fifty thousand had died of starvation and disease. The Spaniards' native allies, particularly the Tlaxcalans and the Texcocans, suffered losses running into the tens of thousands. Several hundred Spaniards were lost, either through capture and sacrifice or on the battlefield.

The great treasure store of Moctezuma, the bulk of which the Spaniards had been forced to leave behind during the *noche triste,* was never found. Cuauhtemoc was tortured by having his feet burned in a vain effort to find the supposedly hidden treasure. Small caches of gold and jewels were found, and the total was reckoned at one hundred thirty thousand gold pesos. The king's fifth, twenty-six thousand pesos, was, ironically, taken by a French pirate while en route to Spain. The share that eventually trickled down to the men was estimated at sixty pesos for a horseman, fifty for a foot soldier — just about the price of a new sword or crossbow.

But before this paltry distribution Cortés, perhaps anticipating the dissatisfaction that would follow, arranged a victory banquet at his headquarters in Coyoacan, one of the few parts of the city that had survived the war more or less intact. A shipment of wine from Spain had been received, also a load of hogs from Cuba. All the soldiers were invited to the feast and so were the five Spanish women who had made their way to Mexico with one or another of the shiploads that had strengthened the invading army.

Bernal Díaz recorded the scene in the first draft of his memoirs: "All the captains and soldiers who had distinguished themselves were invited, but when we went to dine there were not enough tables and chairs for a third of us, which caused great confusion. . . . It would have been better if the banquet had not been given. The plant of Noah [that is, wine] caused some men to do foolish things. Men walked on the tables and afterward couldn't manage to make their way out of the patio. Others said they were going to buy horses with gold saddles. Crossbowmen talked of arrows and quivers of gold. Others rolled down the steps. After the tables were taken away, what ladies there were danced with the younger men loaded down with their cotton paddings. . . . A scene that was completely ridiculous."

Bernal Díaz had second thoughts about his graphic description and later altered it to read: "After the tables were taken away there was great rejoicing and thanks were given to God for His many blessings."

FOURTEEN

New Spain

THE CONQUEST AND destruction of the Aztec empire, which Cortés achieved in the thirty-one months after leaving Cuba, ranked with the great military feats of the ages. Bernal Díaz, who was often critical of his commander, conceded that his accomplishment made his name "as much revered as Caesar's and Pompey's in the time of the Romans, Hannibal's among the Carthaginians, or in our time that of Gonzalo Hernández."

Yet the challenge of the conquest was no greater than the challenge that lay ahead for the man from Extremadura. He had to build a white man's empire out of the wreckage, rubble and ashes of the Indian empire he had destroyed. He intended to build a city that would, he promised, be larger and grander than any other ruled by Charles V. He had to search out the riches of the country and keep a steady stream of gold flowing back across the Atlantic to the coffers of an insatiable fatherland. He had to produce new wealth through industry and agriculture. He had to create a new and submissive society in which indigenes could live — separately and unequally — under the authority of an alien church and foreign king. He had to keep peace with his followers and supporters, many of whom were ruthless rascals, capable of any kind of anarchy, insubordination or violence.

And he had to do all this with, at first, no certainty of the legitimacy of his venture, no assurance that his deeds would be officially recognized and rewarded.

The struggle for recognition by royal authority was an extension of his feud with Diego Velázquez. From the outset Cortés had been trying to make himself responsible directly to the king, circumventing Velázquez just as Velázquez had circumvented Diego Colón. Velázquez had won the first round when, by royal decree of November 13, 1518, he was appointed *adelantado,* or governor, of Yucatán — the only name yet given to the new country. Cortés had executed a clever counterstroke with his ingenious creation of the independent town of Villa Rica de Veracruz. This had been followed quickly by his shipment of Aztec treasures to Spain in the care of Montejo and Puertocarrero, who were to serve both as couriers and pleaders of the case for Cortés.

But there were awesome obstacles. The greatest one was Juan Rodríguez de Fonseca, Bishop of Burgos and head of the Council of the Indies. Fonseca had virtually ruled Spain's New World possessions from the time of their discovery. He had opposed and hindered the work of both Christopher Columbus and his son Diego, and was to do the same for Cortés. On the other hand he favored Diego Velázquez, who married his niece, Petronila de Fonseca (some gossips said she was really the bishop's daughter). On Fonseca's orders, the Mexican treasure transported by Montejo and Puertocarrero — the most eloquent argument in favor of royal recognition of Cortés and his seditious undertaking — was seized and impounded in Seville.

Fonseca struck an even more direct blow in April, 1521. He persuaded Adrian of Utrecht (later to be Pope Adrian), who was serving as regent of Castile during Charles's extended absence in Germany, to issue a warrant for the arrest of

Cortés. Fonseca then delegated Cristóbal de Tapia, overseer of the royal gold foundries in Española, to go to Mexico, seize Cortés and bring him back for trial. Tapia arrived in Veracruz on December 4. Cortés carefully avoided a personal encounter with him, but sent various of his officers to deal with him cordially. And, more important, he sent substantial gifts of gold for Tapia. In the end Tapia accepted the arguments of the Cortés surrogates, pocketed the gold and left.

In July of the following year Charles V returned to Spain and was confronted by the problem of what to do about Cortés. By this time it was clear that Cortés had conquered an important overseas empire. He had also sent a considerable quantity of gold, which Charles badly needed and which Fonseca had managed to tie up in Seville. The conqueror's father, Martín Cortés, Montejo, Puertocarrero and the Duke of Bejar were pleading the case eloquently. Fonseca and other Velázquez partisans presented outrageous charges against Cortés, some true but many pure fiction. Charles referred the matter to a board of inquiry. At Valladolid on October 15, after hearing the board's findings, Charles signed a decree in favor of Cortés, appointing him governor of New Spain, a position he already occupied on a de facto basis. Cortés, the daring soldier and master manipulator, had won both legality and a major victory over Spanish bureaucracy. But bureaucracy was not wholly defeated, for, at the same time and probably at the urging of some of the bureaucrats, Charles signed an additional decree appointing four royal officials to "assist" Cortés in his official responsibilities and, in all likelihood, to oversee him.

It was many months before Cortés learned of his victory. Meanwhile, much had been happening.

Almost before the smoke and death-stench of Tenochtitlan had blown away, Cortés had a multitude of native workers clearing debris, filling the canals, salvaging stone from wrecked buildings. Cuauhtemoc was made to order his people back into the city from their hiding places, to bury or burn the dead, to clear and rebuild the streets, to restore the wrecked aqueduct.

Armies of Indians — some of them the conquered Aztecs but many of them Tlaxcalans and Texcocans who had fought on the winning side — went into the mountains and dragged back huge building stones and great logs of pine, cedar and cypress. The rainy season was over by the time work got under way. The air was filled with dust and the aroma of fresh-cut timber. The boundaries of the city were expanded on made land; where the waters of the lake had once been, new buildings rose on foundations of rubble, human bones and, possibly, some of the gold and jewels the Spaniards had unsuccessfully tried to carry away on the *noche triste*.

The causeways were broadened and reinforced, becoming wide avenues. The wreckage of the great temple was leveled and a cathedral built on the site. The vast square that the temple had faced became a typical Spanish plaza. At one corner of it, the site of Moctezuma's principal residence, a palace was built for Cortés; it was said that in addition to the building stones, seven thousand tree-size cedar beams were used in its construction. Cortés, however, continued for some time to occupy his quarters in Coyoacan.

The new city, Mexico, rose almost as rapidly as Tenochtitlan had disappeared. The central city was Spanish. The Indians were urged to build new homes of their own, but to do so in the suburbs of Tlatelolco or Popotla.

Ixtlilxochitl, the Texcocan prince who had remained loyal to the Spaniards even in the darkest days of the siege when other Indian allies were mysteriously disappearing, was said to have put four hundred thousand of his own subjects to work rebuilding the city and claimed to have rebuilt forty thousand houses of the kind that the city had had before its destruction and one hundred thousand better than before. (Later, when Cortés rewarded Ixtlilxochitl for this and other services by giving him and his descendants three provinces with several scores of towns, Ixtlilxochitl responded that what Cortés proposed to give him already belonged to him and had long belonged to his ancestors.)

It was forced labor of the worst kind. The natives were expected to provide all building materials, most of which they had to carry on their backs or drag along the ground. If they wanted shelter they had to provide it themselves, and also food, which was critically short because of the war's destruction of fields and disruption of transportation. The Indians, accustomed to hard work but in short shifts with frequent rest periods, were made to work nonstop from dawn to dark. Those brought up from the subtropical lowlands suffered from the altitude and cold of Mexico. López de Gómara said, "They worked so hard and ate so little, the pestilence attacked them and an infinite number died."

The native work force was, at first, managed and directed through surviving Indian lords who could command them. This evolved into the *encomienda,* or trusteeship, system which the Spaniards had employed with disastrous results in the Indies. Whole Indian communities were given to individual Spaniards, who were to exact from them tribute in both goods and personal services in return for "protection" and indoctrination in the Catholic faith and loyalty to the Spanish crown. The condition of the Indians in an *encomienda* was, perhaps, a shade better than the outright slavery into which many of the vanquished had been forced. Nor was it much worse than the tribute-paying vassalage most of the Indians had endured under native rulers — a system in which the tribute could be human life.

Spanish colonial administrators, however, aware of the way in which *encomiendas* were destroying the native races, had tried to eradicate them in the Indies as early as 1503. And in 1520 the Spanish crown had declared all Indians to be free vassals of the king. But Spanish conquerors and settlers paid no attention to such orders. It was to be a troublesome and controversial matter for Cortés.

Cortés was, at first, opposed to the use of the *encomienda* in

New Spain — or he at least pretended to be. In his letter to
the king of May 15, 1522, he said: ". . . The natives of these
parts are of much greater intelligence than those of the other
islands; indeed they appeared to us to possess such an understanding as is sufficient for an ordinary citizen to conduct
himself in a civilized country. It seemed to me, therefore, a
serious matter at this time to compel them to serve the Spaniards as the natives of other islands do."

Still, he went on: if *encomiendas* were not established, "the
conquerors and settlers of these parts would not be able to
maintain themselves." Since the Spaniards were incapable of
supporting themselves by anything other than warfare, there
were two alternatives: either outright enslavement of the
natives, which Cortés had done in the past but which he
appeared to find abhorrent, or a share of the tribute in goods
and services due the crown from the king's Indian vassals
being diverted to the resident Spaniards, both as a reward for
past services and as a guarantee of the future security and
stability of this new branch of the empire.

Cortés was, as usual, acting without authority. And he soon
thereafter, without having heard from Spain, began granting
encomiendas. One of the first was given to his headstrong and
frequently troublesome companion in arms, Pedro de Alvarado. The grant, dated August 24, 1522, read: "By the
present writing there is deposited in you, Pedro de Alvarado, a
citizen of the town of Segura de la Frontera, the lords and
natives of . . . [seven towns] to serve you and aid you in
your properties and businesses in conformity with the ordinances which have been promulgated on this matter. This is
done with the obligation to indoctrinate them in the matters
of our Holy Catholic Faith, with all the vigilance and care
possible and necessary."

With communication being slow, it was not until June of
the following year that the Royal Council in Spain responded
to Cortés, ordering him not to make any *encomienda* grants
and to cancel any that had been made. By this time Cortés

had been handing out *encomiendas* in wholesale fashion, characteristically retaining the richest and most promising ones for himself. He later admitted in a letter to the king that he had deliberately ignored and concealed the order. He repeated his argument that the overseas Spaniards could not exist without such a subsidy. And he added that the alternative for the Indians was worse — enslavement by their native lords whose demands were more arduous and inhumane than those of the Spaniards.

The *encomiendas* remained. And Cortés and his heirs were the most powerful *encomenderos* of all, masters of communities by the hundreds and Indians by the thousands. But the initiative that Cortés had exercised without authority and his subsequent disobedience did not go unnoticed in royal circles in Spain.

Consolidation, exploration and exploitation had begun almost as soon as reconstruction of the capital city. Moctezuma's tribute rolls, records of foodstuffs, textiles, gold and jewels regularly received from various parts of the Aztec realm, were found and carefully studied. Parties of Spaniards supported by native troops and porters were sent out to find the most promising areas. The kingdom of Michoacan, to the west of Tenochtitlan, an ancient enemy of the Aztec regime, submitted voluntarily to the Spaniards. By pushing through Michoacan the Spaniards reached the Pacific coast. There, at a place called Zacatula, Cortés soon undertook the building of four ships at the mouth of the Balsas River. Other shipbuilding was begun on the Tehuantepec isthmus, farther to the south. These ships he proposed to use to explore the great South Sea and find a route to the Spice Islands.

Cortés was also preoccupied with the less exciting but equally essential matter of strengthening the economy. Indigenous crops were few — cotton, maize, chiles, amaranth, native fruits, cacao. It was not enough for the dynamic society Cortés envisioned. He sent out parties to find lands suitable for cultivation or grazing. He sent to Spain and the Indies for breeding stock of cattle, swine, sheep, goats, horses and chickens, and for seeds or cuttings for sugar cane, grapevines, wheat and other grains, citrus fruit, peaches, almonds, various vegetables and starts of mulberry, olive and nut trees. Scarce water was channeled into mills to grind grains or into pastures where stock would graze. Plows and other farming implements were sent for; so were experienced farmers and herdsmen. Cortés correctly guessed that Mexico's great variety of topography and climates could accommodate almost any crop known to the Old World.

In addition to his shipbuilding Cortés was also seeing to the manufacture of armaments. His requests to Spain for cannon, powder and shot had either been ignored or overlooked. He sent men scaling the heights of Popocatepetl to dig sulfur from the volcanic crater in order to manufacture gunpowder. Stones were shaped into cannonballs. He melted copper and tin and forged ninety-five artillery pieces. As a special gift for Charles V he ordered a long-barreled culverin cast in silver. It was decorated with an engraving of a phoenix and a dedicatory verse of the conqueror's own composition:

> As this bird was born unique,
> So am I without rival in serving you
> Who are unequalled in the world.

The lavish cannon was sent off to the king along with enough gold bullion to make a shipment worth a total of more than a half million dollars.

Nor was Cortés neglecting the manners, morals and religious tone of the new society he was creating. As he had done so often before, he banned gambling — although the ban was ineffective and he continued to enjoy it himself. Church attendance was compulsory. There were rigid stipulations as to proper dress.

When Cuauhtemoc, the captive king, complained that many Aztec women of rank were being held by Spaniards as either servants or concubines, Cortés authorized a search of the various conquerors' households, promising that those women who wished to return to their fathers, husbands and children would be permitted to do so. Bernal Díaz recorded that many such women were found but that only three of them were willing to return to their former lives. Most of them declared a newfound distaste for the idolatry of their countrymen; anyway, Bernal Díaz noted, many of them were already pregnant by their new masters and mates. Cortés had fathered a son by the cooperative Doña Marina and had various other women, including at least one more daughter of the late Moctezuma, in various stages of pregnancy. The military-political conquest was being followed by a biological conquest that would create a mestizo society.

Those Spaniards who received *encomiendas* and were already married were ordered to bring their wives to New Spain within eighteen months. Those who were unmarried were given the same length of time to find legal mates — various women had by this time begun to arrive from both Spain and the Indies. Among the arrivals was Catalina Xuárez, the wife Cortés had left behind in Cuba almost four years before. She arrived, without invitation or prior announcement, with her mother, her sister, her brother and his wife and children. Cortés was taken by surprise and, according to Bernal Díaz, was not too happy about it. He had been living in a state of sultanic splendor. Nevertheless, he put on a great show of hospitality, installing Catalina in his Coyoacan house. Within three months, however, Catalina was dead. Officially she was said to have died of asthma. It was the first of a number of deaths in which Cortés was later to be accused of murder for the sake of convenience. Catalina's mother charged him with strangling her daughter.

Cortés had firm ideas as to the nature of his new society. He urged the king to send out "old Christian" women: that is, women who were Catholic by descent rather than by conversion, women who had no trace of Moslem or Jewish blood. He also had specifications for the other kinds of immigrants. He

wanted no lawyers, no physicians and, according to Bernal
Díaz, "no scholars or men of letters . . . to throw us into
confusion with their learning, quibbles and books." He did
not want high officials of the Catholic church, whose venality
might be a deterrent to conversion of the natives, but simple,
humble friars, "holy men of good life and example," to attend
to the religious mission.

In response the Franciscan order sent out twelve friars led
by Father Martín de Valencia. The friars insisted on walking
barefoot from the port of Veracruz to the city of Mexico.
Everywhere they were received with great solemnity. Cortés
himself knelt on the ground to kiss their garments. The
natives, called out to witness the welcoming ceremonies, were
deeply impressed by two things. One was the sight of the great
and powerful Cortés humbling himself before anyone. The
other was the obvious poverty and humility of these vicars of
the white man's deity. One of them, Toribio de Benavente,
was nicknamed by the Indians "Motolinía," the poor one. He
was even more devoted than the others to his vow of poverty
and became a notable benefactor of the Indians as well as a
historian of Indian Mexico, writing under the name the Indians had given him.

Among the first tasks of the friars was to collect tithes,
establish monasteries and to educate the sons of Indian nobility in both the Spanish language and the Catholic faith. They
also saw to the destruction of whatever native temples and
idols remained and tried — at first without great success — to
regularize Indian marriages in terms compatible with Catholic
dogma. The missionary work of the friars, first the Franciscans,
later the Dominicans and other orders, was to have a deeper
and more lasting effect on the character of Mexico than any
other single thing in which Cortés had taken the initiative.

Cortés also kept a watchful eye on potential enemies, both
within his own ranks and from without. A successful poacher
himself, he was hypersensitive to being poached on by others.
After the defeat of Narváez, his greatest suspicion was directed

at Francisco de Garay, governor of Jamaica. Garay was one of the wealthiest Spaniards in the Indies. It was said that he used five thousand Indians just to look after his vast herd of pigs.

As early as 1518 Garay, who was territorially ambitious as well as rich, sent out an expedition headed by Alonso Alvarez de Pineda. Pineda cruised the Gulf coast from Florida to the Pánuco River, the northernmost point reached by the Grijalva expedition, which had come from the opposite direction. In the process he had discovered the mouth of a great river which he called the Espíritu Santo — later the Mississippi. At Pánuco Pineda did some trading with the Huasteco natives and secured a little gold — enough to whet further interest on the part of his sponsor. The following year Garay sent out another expedition. Cortés, while at Cempoala preparing for his march inland, was told of the presence of Garay's ships on the coast and captured a shore party from one of them. From these men he learned of Garay's hope of establishing a colony on the mainland. His efforts to entice the remaining members of the expedition to come ashore failed, and he went on with his plans for a march on the Mexican capital. But he did not forget.

Garay sent an agent to Spain and, through the offices of Cortés's enemy Fonseca, obtained authorization to establish a Pánuco settlement. This was in the fall of 1521. Meanwhile, without waiting for authorization, Garay had sent still another expedition to Pánuco — three ships, one hundred fifty men, seven horses, armament and materials for the building of a fortress. The Huastecos, a sturdy and determined people who spoke a language similar to that of the Mayans, attacked them fiercely. One ship sank, and the surviving Spaniards, most of them either ill or wounded, fled down the coast to Veracruz. Here they agreed to join forces with Cortés, who was at Segura de la Frontera preparing for the final siege of Tenochtitlan.

Through his own agents in Spain Cortés heard of Garay's official commission to establish a settlement. Once Tenochtitlan had fallen he began to take countermeasures. His representatives in Spain pressed efforts to cancel or nullify Garay's commission. They claimed that Cortés had already been in negotiation with the Huastecos and that they were ready to become vassals of the king of Spain, which was not true; and that Garay was conspiring with Diego Velázquez to discredit Cortés and the work he was doing in the king's name in New Spain.

But Cortés was not content to let matters rest here. By midsummer, 1522, he was preparing to march on Pánuco and establish a settlement of his own before Garay could do anything further. His departure was delayed, probably by the arrival of his wife, Catalina. But in November, shortly after Catalina's demise, Cortés was off for Pánuco, riding at the head of an army of three hundred Spanish foot soldiers, one hundred fifty horsemen and forty thousand native warriors. He met a hostile force of about seventy thousand Huastecos. The battle occurred on a level plain, where the Spanish cavalry could be used with maximum effect, but the Spaniards lost fifty men and five thousand of their Indian allies, while the Huastecos lost about three times that. It was the first of a series of bloody clashes, and the Spaniards were made nervous by the sight of tanned skins with recognizable bearded faces intact on display in native temples — unfortunate members of Garay's expedition. The Huastecos rejected Cortés's peacemaking efforts; when they could fight no more they fled.

Cortés established a town, Sanesteban del Puerto, with a garrison of one hundred foot soldiers and thirty cavalry under the command of Pedro de Vallejo. Then, after distributing some *encomiendas* to members of the garrison, he returned to Mexico, reaching there in early spring.

Meanwhile Garay was preparing a more ambitious expedition, one which he would lead himself: nine ships, two brigantines, eight hundred fifty men, one hundred forty-four horses and elaborate armaments. With this large force he sailed from Jamaica on June 26, not long after Cortés had returned to Mexico. Cortés was said to have written letters to Garay offering his assistance and cooperation in settling the Pánuco region; and Diego Velázquez, who had come to Jamaica to help Garay mobilize his forces, warned the Jamaican governor of the sort of duplicity he could expect from Cortés.

Garay landed at the Rio de Palmas, north of Pánuco, and marched south along the coast, struggling through swamps and hunting for places where the numerous rivers could be crossed. Finally he came to the new town Sanesteban del Puerto, where Vallejo, the Cortés lieutenant, told him that it would be impossible for the little settlement to support such a large force. Garay was persuaded to march on to a deserted Indian town and to disperse his forces among other Indian settlements. Garay sent a letter to Cortés, complaining of the latter's establishment of a Spanish town in a territory that was, by royal decree, the province of Garay.

Cortés responded vigorously. He sent a force headed by Pedro de Alvarado. Alvarado captured one of the scattered sections of Garay's army, deprived its members of their arms and escorted them to Sanesteban as prisoners of war. Meanwhile, Vallejo, the commander of Sanesteban, had seized Garay's ships and put his own men in command of them. Cortés, at about the same time, had dispatched a second large force to Pánuco headed by Diego de Ocampo and Francisco de las Casas. They informed Garay that Cortés was now officially the governor of New Spain and that a royal edict dated April 24, 1523, had arrived from Spain, and that it ordered Garay *not* to settle at Pánuco but to go elsewhere.

Garay, who must by this time have been totally bewildered, complained that his ships were in no condition for a voyage and that anyway they were in the hands of Vallejo. As an alternative he agreed to go to Mexico and discuss matters with Cortés. This he did, leaving the bulk of his disorganized and leaderless army scattered among various Indian towns of the Pánuco region.

Cortés welcomed Garay to Mexico in November with honors and a great show of cordiality — "such hospitality as I would have shown my brother," Cortés wrote. Garay responded by addressing Cortés as "very magnanimous lord." An alliance was agreed upon, signalized by the betrothal of Catalina Pizarro, a bastard daughter Cortés had sired by a Cuban Indian woman before leaving on his adventure, and Garay's eldest son. It was agreed that Cortés would support Garay in establishing a settlement at Rio de Palmas, where he had originally disembarked. So amicable did their relations become that Garay named Cortés executor of his estate — not knowing that the services of such an executor would become necessary almost immediately.

Cortés and his erstwhile rival attended matins early on Christmas morning and afterward took breakfast together. Garay became ill and died a few days later. Cortés's enemies were to charge that Garay's breakfast of eggs seasoned with herbs had also been laced with poison. López de Gómara, whose source of information was Cortés, said later that Garay had suffered a chill and developed pneumonia. Bernal Díaz said it was pleurisy. Cortés, in a letter to the king, said that the cause of Garay's death was the rebellion of natives in Pánuco that had broken out after Garay's departure and spread like wildfire, causing heavy losses in Spanish lives: "The *adelantado* was so stricken by this news, because he believed he was the cause of it and because he had left one of his sons and all that he had brought with him in that province, that he fell ill of his grief, and of this sickness passed from this life within the space of three days."

The aftermath of the Garay affair was to be another dark and puzzling chapter of the conquest. The men Garay had left behind apparently had been marauding around the countryside, and the natives had retaliated in bloody fashion, killing several hundred of Garay's men and forty-odd of the little garrison at Sanesteban.

Cortés dispatched Sandoval with fifty horsemen, one hundred crossbowmen and musketeers and an Indian force of thirty thousand. Sandoval made short work of it. He defeated the Indians in the field and rounded up some four hundred of their chieftains. Cortés, who did not bother to identify Sandoval except as "a Spanish captain," reported to the king that "the chieftains were burnt in punishment, for they confessed to being the instigators of the war, and each one to having killed or had a part in killing Spaniards."

The Pánuco incident had overtones of duplicity. García del Pilar, a Spaniard who served as interpreter with Sandoval, later said that the chieftains were not captured but, instead, after Spanish battlefield victories, were summoned together to hear some announcements. When the chieftains came in they were imprisoned. Each was asked how many Spaniards he had killed and each responded truthfully. Sandoval then ordered them to be burned. While they were being tied to stakes they appeared puzzled and asked Pilar why they were being punished. Pilar explained that it was for having killed Spaniards, which they had admitted. But, said the Huastecos, this was what they had been instructed to do by the Spaniards' Indian allies. It was, they had been told, what Malinche, or Cortés, had ordered.

The charges of double-dealing with both Garay and the unfortunate Huasteco chieftains were only two of many accusations that were being directed at Cortés. Among them:

• That he had recovered and hidden Moctezuma's treasure for himself, sharing it with neither his king nor his men.

• That he had a personal fortune of three hundred million or more *castellanos*.

• That he had snubbed and neglected old comrades and favored newer, more powerful and potentially helpful friends, especially if they came from Medellín; that in his distribution of *encomiendas* he had kept almost forty of the better ones for himself, a territory as large as Andalusia.

• That he had murdered his wife and his rival, Garay.

• That the ships he was building ostensibly for the Spice Islands trade were really for thep urpose of carrying his personal treasure to France, out of the reach of Charles V.

• That he was tyrannical, that he was disloyal to his king, that he was planning an independent Mexico in which he would be the absolute ruler.

That Cortés was self-serving and secretive was undeniable. But the suggestion of disloyalty to his king was baseless. Cristóbal Pérez Herrera, an aide of Garay's who was sent to Spain to attend to some of his late leader's business, had every reason to dislike and distrust Cortés. Still, he told the historian Peter Martyr that neither he nor anyone else "[had] ever observed the smallest sign of treason in him." Pérez Herrera also added a description of Cortés at the time. "Cortés usually dresses in black silk; his attitude is not proud, except that he likes to be surrounded by a large number of servants, secretaries, valets, ushers, chaplains, treasurers, and all such as usually accompany a great sovereign. Wherever he goes he takes with him four *caciques* on horseback. The magistrates of the town and the soldiers armed with maces to exercise justice precede him. As he passes everyone prostrates himself, according to an old custom. He accepts salutations affably, and prefers the title of *adelantado* to that of *gobernador*, both dignities having been conferred upon him by the emperor."

Cortés was at the very pinnacle of his career. He was dining with trumpets, just as he had long ago dreamed he would. It was a climax of grandeur, glory and power that he would never quite achieve again.

FIFTEEN

Honduras

FRAY TORIBIO DE BENAVENTE, the humble Franciscan who adopted the Indian name of Motolinía, or the poor one, listed what he considered the plagues of New Spain, measuring them by their adverse effect on the natives. They included the wholesale slaughter during the conquest, the famine that immediately followed it, smallpox, the Spanish-aggravated cruelty of Indian to Indian, the excessive tribute and personal services demanded by the Spaniards, the gold mines and the enslavement of natives either to work in the mines or to transport freight, and finally, the dissensions among the Spaniards themselves.

Dissension was built into the character of the highly individualistic and strong-willed Spaniards. The success that Cortés had achieved in Mexico had been repeatedly threatened by such dissension. Often it was intensified by Cortés's inequitable distribution of the spoils of war. But even more often it sprang from the long-smoldering feud between Cortés and Diego Velázquez.

When the crown finally ruled in favor of Cortés and against Velázquez, the news of the decision was a crushing blow to

Velázquez, and one which he did not long survive. However, like some wild creatures that are most dangerous when they have been mortally wounded, Velázquez was to make one more strike against his old enemy, and it was to be a devastating one. After settlement of the Pánuco-Garay affair Cortés's realm measured perhaps a quarter of a million square miles. Little of it was known. There were still stubborn pockets of Indian resistance. And there were elusive but tantalizing rumors of gold as yet unfound. Cortés sent out parties of Spaniards in all directions to discover and exploit whatever might be there.

There was also a specific order from Charles V, dated June 26, 1523, to search for a strait connecting the Atlantic and Pacific. Spain needed easy access to the riches of the Orient. Cortés was building ships for exploration both north and south along the Pacific coast, and there were plans to explore the Gulf coast north and east from Pánuco to Florida and perhaps beyond. Trusted captains were also sent to pacify the natives and establish Spanish towns in the areas of Jalisco and Colima, in southern Veracruz, and farther to the south, in Oaxaca, Chiapas and Guatemala.

One of the most alluring areas was the coast south of Yucatán. Balboa had already demonstrated the narrowness of the isthmus separating the two seas in Darien (Panama), and it was thought that if the strait existed it might well be found along this coast. There was also the enticing story that native fishermen in this region used nets weighted with gold. For Cortés there was also the danger that someone else might get there first. Pedro Arias de Avila was pushing north from Darien, and Gil González de Avila was operating in what is now Central America under the orders of the colonial authorities in Española.

Accordingly Cortés chose one of his most able captains, Cristóbal de Olid, to explore the territory below Yucatán, known vaguely as Las Hibueras — now Honduras. Olid had served valiantly as a key commander in the siege of Tenochtitlan and, more recently, had commanded expeditions into

Michoacan and Colima. Olid, said Bernal Díaz, "had a good person and countenance, a cleft in his under lip, and his voice was rough and fierce. . . . [He had] many good qualities, being sincere, and for a long time [was] much attached to Cortés."

On January 11, 1524, Olid sailed from Veracruz with four-hundred-odd soldiers abroad five ships and one brigantine, bound for Honduras by way of Cuba, where he was to pick up additional colonizing supplies, including more horses. Many years later Bernal Díaz was to say of Olid, perhaps with the wisdom of hindsight, that "the ambition of governing and dislike of being governed perverted his mind."

Diego Velázquez, a bitter man who had only a short time left to live, apparently sensed this in Olid. He suggested. that Olid set aside his allegiance to Cortés and declare himself an independent conquistador in Honduras, just as Cortés had done in Mexico. This Olid promptly proceeded to do after reaching Honduras in May, 1524, and establishing the settlement of Triunfo de la Cruz.

Cortés might have long remained ignorant of Olid's rebellion had it not been for the royal officials appointed by the crown to "assist" him as governor of New Spain. One of them, the royal agent Gonzalo de Salazar, came through Cuba at about the time Olid and Velázquez were conferring. He learned of the plot and informed Cortés of it when he reached Mexico. Cortés immediately ordered a kinsman, Francisco de las Casas, to take a strong force and go to Honduras, punish Olid and restore order.

If Cortés had been content to let matters rest here and devote all his energies to administration all might have been well. But Cortés was both bored and restless. Although he was an able administrator, even a gifted and farsighted one, his heart was not in it. The royal officials — the agent Salazar, the treasurer Alonso de Estrada, the accountant Rodrigo de Albornoz and the overseer Peralmíndez Chirinos — were interfering in matters that he had, until then, handled without consulting anyone. Encroaching bureaucracy made him long for a field command, to exercise his power and authority unchallenged, the sort of thing at which he excelled. Adding to his restlessness was a broken arm, suffered in a fall from a horse. He had been somewhat restricted in his movements and was ill at ease. "It seemed to me," he wrote to the king, "that I had for a long time now lain idle and attempted no new thing in Your Majesty's service on account of the wound in my arm; and although that was not yet healed, I determined to engage in some undertaking." His destination: Honduras. It was one of the most fateful decisions of a brilliant career — and one of the most pointless.

There had not been enough time for him to know how or whether Francisco de las Casas had dealt with the Olid problem. Prudence would have dictated waiting to hear if the mission had been accomplished. But he was impatient.

He was equally injudicious in his choice of men to govern in his absence. He delegated his authority to two of the four royal officials, the treasurer Estrada and the accountant Albornoz, with judicial authority to be exercised by Alonso de Zuazo, a recently arrived lawyer. He had reason to distrust both Estrada and Albornoz. Estrada was a middle-aged man with a distinguished civil and military career behind him, but he was still ambitious. He also claimed proudly to be a natural son of the Catholic king, Ferdinand. Albornoz, for his part, had quarreled with Cortés soon after his arrival because the latter failed to give him Indians and specifically refused him the daughter of a lord of Texcoco. Albornoz had a weapon of retaliation: he had brought with him a code with which to communicate secretly with the Council of the Indies.

Canny as he was, Cortés could not have been without his own suspicion of Estrada and Albornoz. That he appointed them anyway was a mark of his supreme self-confidence. To soothe the feelings of the other two royal officials, Chirinos and Salazar, he decided to take them with him on his Honduras venture.

Before setting out from Mexico in October, 1524, Cortés composed another letter to the king. In it he committed an indiscretion that would weigh heavily against him in the future. He had denounced Diego Velázquez in earlier letters. Now he did so more vigorously and specifically. Referring to the Velázquez-Olid conspiracy, he said: "This seemed such an ugly business and such a disservice to Your Majesty that I can scarcely believe it; on the other hand, knowing the cunning which Diego Velázquez had always practiced against me to harm me and hinder my services, I do believe it. . . . I am of a mind to send for the aforementioned Diego Velázquez and arrest him and send him to Your Majesty; for by cutting out the root of all these evils, which he is, all the branches will wither and I may more freely carry out those services which I have begun and those which I am planning."

Cortés's threat to assume authority that could only properly be exercised by the king was bound to raise alarm in Spain, both in the Council of the Indies, already suspicious of Cortés, and in the king himself.

There was further evidence of Cortés's grandiose notions in his preparations for his expedition against Olid. His entourage included several hundred foot soldiers, about half as many horsemen, a majordomo, two masters of the household, a butler, a chamberlain, two stewards (one of them in charge of gold and silver plate), a doctor, a surgeon, various pages, including two lance pages, eight grooms, two falconers, five musicians, an acrobat, a magician-puppeteer, a string of mules to carry freight, a herd of swine to provide food en route and several thousand Indian porters carrying huge loads of powder, horseshoes, iron tools, tents, beads and other trade items.

Although pleased at the prospect of exercising power in the absence of Cortés, the royal officials nevertheless were apprehensive about a possibly long absence. To appease them Cortés insisted that he planned to go no farther than the province of Coatzacoalcos, on the lower Gulf coast of Veracruz, where, two years earlier, Sandoval had established the Spanish town of Espíritu Santo.

At Espíritu Santo Cortés received word that his deputies in Mexico were quarreling and had actually drawn their swords against each other in a disagreement over the appointment of a constable. He thereupon dispatched the two royal officials who had accompanied him, Chirinos and Salazar, with instructions to make peace between their two colleagues. If this could not be done they were instructed to ally themselves with Zuazo and take over the government. Again Cortés was recklessly depending on his personal authority and the stability of the government he had created. He was also compounding his original error in leaving the city, for, as he learned later, Chirinos and Salazar only made matters worse.

Cortés also attended to a more personal matter. He arranged and supervised a marriage between his faithful interpreter, diplomatic adviser and mistress, Marina, and Juan de Jaramillo. Jaramillo was drunk at the time, and many of the Spaniards were secretly critical of Cortés for his treatment of the woman who had not only made his conquest possible but had also borne him at least one child. Despite her marriage Marina continued to serve Cortés as an interpreter.

Finally Cortés began his march. Why he did not make the trip by sea is something of a mystery. He had ships available and pilots who knew these waters. Instead he elected to go overland across virtually impassable country.

First there were broad rivers and tumultuous streams separated by vast marshes. In one hundred-mile stretch the Indian bearers — and those few Spaniards willing to help them — had to build fifty bridges, one of them nine hundred paces long: well over a half-mile. Trees had to be felled, trimmed, dragged to the site, sunk in riverbed or marsh, braced with lighter timbers tied together by vines. It was a disaster from the outset. Spaniards, Indians and horses were swept away in the torrents or sank out of sight in the swamps. Baggage was lost, food supplies were exhausted, and the local natives on whom Cortés had depended for food in the earlier operations deserted their towns and villages when the Spaniards approached. After struggling through the lowlands, the Spaniards encountered deep forests and chains of rugged mountains. It took twelve days for the weary army to traverse a twenty-four-mile-long pass in the Sierra de Pedernales; sixty-eight of the horses lost their footing and plunged into the chasms, and almost all the rest were crippled by either falls or damaged hooves.

As a security measure Cortés had taken with him on the expedition Cuauhtemoc, the captive Aztec king, Tetlepanquetzal, the lord of Tacuba, Coanacoxtzin, the deposed ruler of Texcoco, and various other native potentates who might have been capable of beginning an insurrection in his absence. Cortés knew better than anyone else the vulnerable state in which he had left Mexico, and the danger must have preyed on his mind.

Cortés later reported that while in Izancanac, capital of the province of Acalan, he was advised by a native informer that Cuauhtemoc and the other hostage chieftains were plotting to kill him and all the Spaniards in his company, after which they would stir up a countrywide insurrection and reinstate themselves in power.

"I seized all those lords and had them imprisoned separately," Cortés wrote. "I then asked each of them about the plot, pretending to each that one of the others had revealed it to me. . . . They were thus forced to confess that it was true. . . . These two [Cuauhtemoc and Tetlepanquetzal] were hanged." The others he let go. "I have left their cases open so that they may be punished if they ever relapse, but they are so frightened that I do not think they will, for as they have never discovered from whom I learnt of their plot, they believe it was done by some magic art, and that nothing can be concealed from me." He added that the paroled chieftains thought their conspiracy had been magically revealed by the mariner's compass with which Cortés charted his route.

Some versions of the event say that the two ex-kings were first beheaded and then hanged by their feet. Whatever the method, Cuauhtemoc, before dying, made a little speech to Cortés — although Cortés does not mention it. According to Bernal Díaz, Cuauhtemoc said: "Malintzin! Now I find in what your false words and promises have ended — in my death. Better that I had fallen by my own hands than trust myself in your power in my city of Mexico. Why do you thus unjustly take my life? May God demand of you this innocent blood!" And on his own Bernal Díaz added: "Thus ended the lives of these two great men, and I must say like good Christians, and for Indians, most piously, and I heartily pitied Cuauhtemoc and his cousin . . . and I also declare that they suffered their deaths most undeservingly, and so it appeared to us all, amongst whom there was but one opinion on the subject: that it was a most unjust and cruel sentence."

The execution took place in late February or early March, 1525. Either because of conscience or because of fear of retaliation by the small army of Mexican porters and servants — although they were so near starvation there was little to fear — Cortés was irritable and morose. Moving on to another town Cortés took possession of the principal temple as his quarters after promising the priests he would not touch or harm their idols. He insisted, however, on erecting a wooden cross near the temple. Still he was uneasy, and, according to Bernal Díaz, "very ill-tempered and sad . . . vexed by the difficulties and misfortunes which had attended his march, and his conscience upbraided him with the death of the unfortunate Cuauhtemoc. He was so distracted by these thoughts that he could not rest in his bed at night, and getting up to walk about as a relief from his anxieties, he went into a large apartment where some of the idols were worshipped. Here he missed his way and fell from the height of twelve feet to the ground, receiving a desperate wound and contusions in his head. This hurt he tried to conceal, keeping his suffering to himself and getting his hurts cured as well as he could."

But bad conscience, depression, physical injury and the incredible hardships of the journey did not halt the expedition. Finally, after a march of more than a thousand miles, the exhausted, hungry and ragged men reached the settlement that had been established by Gil González de Avila near the south shore of the Gulf of Honduras. Here they learned the news that Olid, the man whose disloyalty was the reason for the entire excursion, was long dead. Olid had made prisoners of both Gil González de Avila and, somewhat later, Francisco de las Casas, the kinsman of Cortés whom the latter had sent against Olid. But Olid had mistakenly treated the two more as guests than as prisoners and had allowed them ample liberty. The two prisoners often joked with Olid that they would one day kill him. And one day they did, attacking him first with table knives and later beheading him. Then the two had set off for Mexico, marching overland by way of Guatemala, which by now was controlled by Pedro de Alvarado.

Whatever reflections Cortés may have had about the uselessness of his extraordinary march to Honduras were overshadowed by grim news from Mexico. A ship had arrived from Española. On it was a letter to Cortés from Zuazo, his deputy as chief justice in Mexico. Salazar and Chirinos, the royal officials Cortés had sent back to Mexico from Espíritu Santo, had outrageously overstepped their authority. They had reported that Cortés was dead. They had deposed Estrada and Albornoz, imprisoned them and declared themselves governors in their place. They had seized Cortés's property and had executed his cousin Rodrigo de Paz, whom Cortés had left behind as his majordomo. They had arrested Zuazo, the acting chief justice, and sent him in chains to Española. The wife of one of Cortés's companions, who openly declared she did not believe the story that Cortés was dead, was seized and publicly whipped as a witch. On all sides Indians, sensing the disarray of the Spanish establishment, were rising in rebellion. In short, it appeared that all that Cortés had achieved in almost seven years of struggle was rapidly disintegrating into chaos.

Cortés brooded over the bad news, held special masses and religious processions, apparently hoping for divine guidance. Although he was no stranger to discouragement and disaster, he seemed to be at the lowest point of his career and for the first time appeared hesitant and uncertain. He was wasted by injuries and disease, probably malaria. His air was one of deepest despondency, and Bernal Díaz ventured the guess that he expected soon to die; he had ordered made a habit of the Franciscan order, in which he hoped to be interred.

But in the end he decided that he must try to return to Mexico. He ordered Sandoval to march overland by way of Guatemala while he, Cortés, boarded ship in Honduras. He set sail three times and three times was turned back by storms at sea. Cortés thought this might be an expression of divine will. He ordered another series of masses and processions, and sent one of his aides, Martín Dorantes, as a messenger in his stead. Dorantes was to convey the information that Cortés was indeed alive, and would return.

Cortés talked of an expedition into Nicaragua. Before this could happen his cousin Fray Diego Altamirano arrived in Honduras with even more harrowing stories of the anarchy in Mexico. Cortés thought at first of marching back to Mexico by way of Guatemala, but Altamirano urged that he go by sea, which would be quicker, and it was vital that he reach Mexico as soon as possible. He set sail, had a prosperous voyage to Cuba, where he rested a few days, and then sailed on to Veracruz. He landed there at night — either because of adverse weather conditions or, possibly, for fear that he might meet with armed opposition. It was May 24, 1526 — nineteen months since he had left on his disastrous excursion. He walked to the nearby town of Medellín and entered the church to pray. At dawn the Spanish settlers of the town heard that there were strange men in their church and came to see. Cortés was gaunt and wasted. His usually trim figure was misshapen, his belly distended. At first the people of Medellín did not recognize him. But when they did, they welcomed him with enthusiasm and relief.

Dorantes, the messenger Cortés had sent ahead to Mexico, had arrived at the capital four months earlier. Dressed in rags, with his commander's messages tied around his waist, he had made his way to the monastery of San Francisco, where a number of men still loyal to Cortés had taken asylum. Cortés had sent word with Dorantes that Francisco de las Casas and Pedro de Alvarado should take over as deputy governors. Unfortunately, Alvarado was in Guatemala, and Las Casas, immediately upon his return to Mexico, had been arrested for the murder of Olid, thrown in chains and shipped to Spain. Cortés's orders had provided that if neither Alvarado nor Las Casas were available, the government should once more be taken over by the treasurer Estrada and the accountant Albornoz, the officials Cortés had originally left in charge. The usurpers Chirinos and Salazar were expelled from office; Chirinos was away from the city fighting Indians, but Salazar was imprisoned in a cage. A new regime was established that was at least outwardly loyal to Cortés.

After resting briefly at Medellín, Cortés set off for the capital. At all points along the route he was enthusiastically greeted by large crowds of Indians, many of them bearing gifts of food, gold, jewels, featherwork and cotton goods. Flowers were scattered in his path. However much they might have suffered under the direct rule of Cortés, they had suffered vastly more under his deputies during the year and a half he had been gone. When he reached the capital he was welcomed with great solemnity by Spaniards and Indians alike. Albornoz met him on the road near Texcoco, and Estrada led the welcoming party in the capital. The Indians lighted bonfires, played fifes and conch shell trumpets and danced in the streets. Cortés secluded himself in the monastery of San Francisco in order, he said, to give thanks to God. But he also must have been doing some careful calculating.

Albornoz, despite his outward show of loyalty, had sent a series of coded messages to the Council of the Indies regarding Cortés and New Spain. One member of the council, the historian Peter Martyr, later wrote that because of these secret messages, "written against Cortés's mad designs, consuming avarice and partially revealed tyranny," the council was "not without suspicion" of the conqueror of New Spain. The conqueror's long absence on his futile adventure in Honduras and the resulting anarchy in Mexico had aggravated these suspicions, as had the conqueror's threat to take royal authority in his own hands and arrest Diego Velázquez. In addition, Charles V, always pressed for funds to support his military and political ventures in Europe, was agitated by reports that Cortés was keeping more treasure than he was sending to Spain and by the rumors that he would declare himself independent of the mother country. "The reports concerning Cortés and his extreme ability in the art of deceiving and corrupting people are contradictory," wrote Peter Martyr. "It seems certain that he possesses such quantities of gold, pearls and silver as have never before been heard of. They are brought to him by a back door of his enormous residence on the shoulders of slaves . . . we are working secretly to devise certain preventive measures."

The council's deliberations on Cortés were going on while there was still some uncertainty whether Cortés had survived his mission to Honduras. It was decided to send to New Spain a distinguished young lawyer, Luis Ponce de León, to conduct, in the name of the king, a *residencia,* or investigation, into the affairs of Cortés, whose authority as governor would meanwhile be suspended. If Cortés was alive, Ponce was to flatter him and reinforce his loyalty to the king. Whether Cortés was alive or dead, Ponce would assume, for the time being, most of the power Cortés had been exercising. Ponce sailed from Seville in a fleet of twenty-two ships in February, 1526. He stopped for several months in Española to look into the records of colonial affairs, and arrived in the Mexican capital early in July, only shortly after Cortés himself had returned.

Cortés, in a letter to the king, said that he was "much pleased" by the arrival of Ponce, and in simulated gratitude added, "I kiss one hundred thousand times the Royal feet." He ordered a banquet held in Ponce's honor in Ixtapalapa when the royal agent reached that city. Many of the party were feeling unwell after their long sea trip, and they were served large portions of custard and curds. After the meal many of them fell ill — and stories of deliberate poisoning began to circulate.

Two days later the king's agent summoned all the officials together in Mexico City, read to them the royal messages and relieved them of the staves they carried as badges of office. All, Cortés included, kissed the king's documents and placed them on their heads as a sign of acceptance and submission. After this Ponce returned all the staves except that of Cortés, the governor, saying in a calm, polite tone, "I shall retain for myself the staff of the Lord Governor." A public crier then announced the official investigation and invited all who had complaints or grievances against Cortés to come forward for a hearing.

Before the proceedings were well under way, Ponce fell ill with a high fever and accompanying drowsiness. Nine days later he was dead, apparently of an ailment that had already killed many of his party, some while at sea, others after arrival in Mexico. But once more it was rumored that Cortés had contrived to eliminate a threat to his power — this time by having arsenic put in the custard and curds served at the Ixtapalapa banquet.

Cortés, who had weathered such gossip before, paid little attention to it now. He was, for the moment, more concerned with trying to recover the property and belongings that had been seized during his absence. On the presumption, or perhaps in the hope, that Cortés was dead, Chirinos and Salazar had ordered the seizure of his property to pay for a sumptuous state funeral which they had ordered held, and also to pay for perpetual masses in memory of the conqueror.

During his idle hours Cortés amused himself with Doña Isabel, one of Moctezuma's daughters, whom the unhappy king had left to Cortés as a ward. They spent much time together in the rural retreat of Cocoyoc. Later in the year, when Doña Isabel was married to Alonso de Grado, she was pregnant by Cortés, and Cortés, in gratitude, proposed to give her the town of Tacuba as an *encomienda*.

Before dying, Ponce had designated as his successor Marcos de Aguilar, who had come with him from Spain. Aguilar was elderly, frail and syphilitic. His stomach was so delicate that he could take nourishment only from goat's milk or, Bernal Díaz insisted, being breast-fed by a woman from Castile.

Cortés's partisans urged that he resume the governorship himself, disregarding the succession of Aguilar, which they insisted was illegal. Cortés not only refused to do it, but urged Aguilar to continue with the investigation Ponce de León had begun. Aguilar tried to do so, but his health worsened, and within seven months he too was dead. It was difficult to suggest that a man who lived on such a restricted diet could have been poisoned. Nevertheless there were rumors that Cortés had sent Aguilar a gift of Flemish bacon and that Aguilar gave the bacon to a servant, who died shortly thereafter.

Before dying, Aguilar designated the treasurer Estrada as his successor. Estrada, who had at first been only one of four royal officials, was now in a position of sole authority, and he enjoyed the exercise of it, particularly anything inimical to Cortés. When several Cortés retainers stabbed a man in a fight, one of them was arrested, imprisoned, and had his hand cut off on Estrada's orders. He also ordered Cortés into exile to prevent any effort to liberate the prisoner. Cortés prepared to leave the city. But on the same day there arrived in Mexico a Dominican friar who had just been named Bishop of Tlaxcala. Hearing of the Cortés-Estrada feud, he intervened, brought the two men together and persuaded them to reconcile their differences.

By this time Cortés had decided that he must go to Spain. There were, as López de Gómara pointed out, several reasons: "One was to find a wife, for he had children [although he had had none by his first wife] and was getting old; another was to appear before the King in person and give him a full account of the vast territory and the many peoples he had conquered and partly converted, and inform him by word of mouth of the quarreling and dissension among the Spaniards in Mexico, for he suspected that the King had not heard the truth; another was to demand of him rewards commensurate with his merits and services and a title that would set him above the others. He had, moreover, a number of profitable suggestions."

SIXTEEN

Spain

IN THE SPRING of 1528 Cortés had suffered many and severe drains on his fortune. Still, he was a vastly wealthy man — probably worth more than a million dollars. And he was to spare no expense in making his trip to Spain a display of virtually oriental splendor. He would deal with his king from a position of obvious power.

Cortés had learned to use regal manners, openhanded generosity and grand gestures as psychological weapons in dealing not only with the aboriginal peoples of America but also with his Spanish companions.

Two ships were chartered, outfitted and provisioned at Veracruz. Any Spanish settlers in New Spain who wished to do so were invited to accompany him. So were several score of Indians, among them three sons of Moctezuma and one son of Maxixca, the old chief of Tlaxcala who had been both friend and benefactor to Cortés. The Indian company included albinos, hunchbacks and sufferers from other abnormalities: there was, in sixteenth century Europe, a strange fascination with the odd and freakish. There were also Indian performers. Some were adept at the Indian ball game in which a large, solid rubber ball was propelled by being struck with the buttocks. There were jugglers and acrobats. Some performed on a high pole, around which they revolved rapidly at the end of ropes, seeming to fly. Others lay on their backs and juggled whirling wooden logs with their feet.

Cortés also took a vast collection of birds and animals that would be unfamiliar and exotic in Spanish eyes, including opossums, armadillos, albatrosses and "tigers" — probably jaguars or ocelots. There were tubs of ambergris and balsam. There was also a vast assortment of examples of Indian handicraft and artistry — featherwork, weaving, fans, shields, decorative plumes, precious stones that had been worked into intricate forms or polished to serve as mirrors. Most important, there was a cargo of twenty thousand pesos of fine gold (about $120,000 worth), a lesser amount of low-grade gold and fifteen hundred marks' worth of silver.

On his person Cortés carried a pouch containing five fabulous "emeralds," elaborately worked by Mexican lapidaries. They may have been true emeralds, since native traders sometimes brought goods from as far afield as northern South America, where there are deposits of emeralds; or they may have been nothing more than splendid examples of jadeite found in Mexico. Whatever the case, the workmanship was intricate and beautiful. One had been carved to resemble a rose, another a horn, another a fish with golden eyes. Another was a gold-trimmed bell with a pearl clapper, engraved with the words "Blessed be he who made thee." And finally there was a green stone cup decorated with gold chains and a large pearl, with the legend "Among those born of woman there is none better." Of all the treasures and curiosities that Cortés brought with him, none was to excite as much envy and comment as the five green gems. A merchant from Genoa later tried to buy one of the jewels from Cortés for forty thousand ducats — about two hundred thousand dollars — intending to resell it to the Sultan of Turkey for a much higher price. Cortés spurned the offer. The stones were important trappings in the role he was playing: a man who had left Spain twenty-four years before as a penniless young scapegrace, who was now returning a hero and conqueror, a man who had added a continent to Spain's empire, who had tamed and Christianized a barbaric people and, most important, a man who had poured a steady stream of gold into the chronically empty purse of Charles V. "In short," said López de Gómara, "he traveled as a great lord."

The ships sailed from Veracruz on March 17, 1528, and reached Palos forty-one days later. Although there had been no preparation for an official welcoming ceremony, the arrival caused more excitement than Spain had seen since the return of Columbus from his first voyage of discovery. The conqueror's face was pale, his thin hair streaked with gray, his manner grave. But his clothing and jewelry were elegant. He and the veterans who accompanied him wore heavy gold chains, and the Indians who accompanied them were dressed in gaudy splendor. Since there were no accommodations in Palos for such a large company they took temporary quarters in the nearby monastery of La Rábida — the same place in which Columbus had taken shelter.

But although the excitement and admiration were gratifying, there also was an element of sadness in the occasion for Cortés. His father, Martín, who had worked so hard pleading his son's case in official circles, had died. And almost immediately upon their return, death claimed Sandoval, most trusted and loved of the various men who had helped Cortés achieve the conquest. Ailing upon arrival, young Sandoval (he was only thirty and Cortés habitually addressed him as *hijo,* son) had taken shelter in the home of a rope-maker at Niebla, a village on the Tinto River just outside of Palos, where he would be more comfortable than in the sparsely furnished monastery. Helpless with fever, Sandoval watched from his bed while his rope-maker host stole thirteen gold bars from his sea chest. Sandoval lived only long enough to report the theft to his companions. With what must have been a heavy heart

Cortés ordered his young friend buried at La Rábida. Sandoval had served his leader selflessly and with astonishing courage and skill. For him to be cut down in his youth without achieving the wealth and honors he deserved may have given Cortés a sense of foreboding.

Cortés sent word to the king that he had arrived and awaited his pleasure. Meanwhile he visited Seville, where he was a guest of the powerful Duke of Medina Sidonia. From here, with handsome horses the duke had presented to him as a gift, he moved on to the shrine of Guadalupe, a favorite retreat for both the royalty and nobility of Spain. Here Cortés proposed to spend nine days in devotions. Here also he met Doña María de Mendoza, wife of Francisco de los Cobos, the king's secretary and one of the most powerful men in Spain, and Doña María's unmarried and very beautiful younger sister, Doña Francisca. The presence of two attractive and very well connected ladies, both of whom were clearly impressed with this glamorous and wealthy man, moved him to display his greatest charms. He gave them trinkets of gold and silver and panaches of green plumes trimmed with gold. He had his Indian dancers, jugglers and acrobats perform for them. When one of Doña Francisca's mules became disabled, he generously bought her two new and better ones. Many observers felt that with all these attentions he was raising expectations on the part of Doña Francisca that he had no intention of satisfying. If true, it may have contributed to the animosity that Francisco de los Cobos was later to display toward Cortés. Had the widower Cortés taken advantage of his opportunity and married into this powerful family, he might have saved himself much future grief and disappointment.

But this was impossible. While still in Mexico he had become betrothed to a young Spanish woman whom he had never seen. This was arranged through his father and other legal representatives and friends in Spain. Two of his most powerful advocates at court were the Count of Aguilar and the Duke of Bejar. The young woman selected as suitable for the conqueror of New Spain was Doña Juana de Zúñiga, daughter of Aguilar and niece of Bejar. Although he had yet to meet Doña Juana, Cortés was — aside from some apparently mild flirtations — honoring his commitment. The five green jewels he carried with him were intended as gifts for his bride.

First, however, he had to transact his business with the king. The court, forever on the move, had just shifted from Madrid to Toledo, and it was at Toledo that Cortés caught up with his monarch.

Charles was still a young man, fifteen years the junior of Cortés, and a somewhat peculiar one. He had straight red hair, pale skin, a broad forehead and bulging eyes that, one observer thought, appeared to have been stuck on as an afterthought, and with these eyes he looked at the world with what appeared to be suspicion. He had an aggravated example of the jutting Hapsburg jaw — so aggravated that speech was difficult and mastication almost impossible; the latter difficulty coupled with a gluttonous appetite (he had a particular passion for anchovies) made him susceptible to numerous ailments. Perhaps because of his difficulty with speech, he maintained silence much of the time and a reserve that bordered on melancholy. In a technical sense he did not have a clear and exclusive title to the throne but, instead, shared it with his mother, the mad Joanna, whom he kept locked up in an asylum. She was not to die until 1555.

But these were details that in no way detracted from his royal eminence. Contemporaries described him as the greatest Christian monarch since Charlemagne. He was, among other things, "emperor-elect" of the Holy Roman Empire, having been chosen for this post by well-bribed electors in 1519. Cortés and others addressed him in correspondence as "Very High, Very Powerful and Most Excellent Prince, Very Catholic and Invincible Emperor, King and Lord," or, somewhat more simply, "Caesar" or "Your Caesarian Majesty."

Charles's ambitions were great and his expenses heavy. His

political maneuvers and frequent military operations (at the end of his career he bragged that he had fought forty campaigns in Europe and Africa) were a constant drain on Spain's limited resources; the expense plus his prolonged absences from the country contributed to two bloody civil insurrections. In his early years his own foreignness and his reliance on foreign advisers, principally Flemish, did not increase his popularity.

Cortés, in preparing to meet his king at Toledo, must have had misgivings. He had spared nothing in obtaining a vast extension of Charles's empire; he had transmitted to him fabulous treasures in gold and jewels; he had spent the best years of his life and much of his blood in doing all this; and he had been cruelly rewarded by being relieved of the governorship that he thought was rightfully his.

Charles, for his part, must have had doubts about Cortés, too. This was a man who had been insubordinate, reckless, sly and ruthless and one who was quite capable of rebelling in the interest of his own complete autonomy and independence.

But their encounter, in the fall of 1528, went well. Despite what his enemies had said of him, Cortés was genuinely loyal and subservient to the king. And the king, for his part, was impressed with Cortés's clear thinking and direct way of speaking, his obvious knowledge of Spain's Indian subjects in the New World and his sound opinions in matters of administration.

Cortés conveyed to the king his genuine concern for the welfare of the Indians of New Spain and his conviction that the native races were one of the principal resources of the overseas realm, a resource that must not be plundered and destroyed but, instead, protected and cultivated. That Charles was receptive to such arguments was evident in orders he gave regarding the visiting Indians. They must be returned safely to New Spain. And they must be dressed according to royal specifications and at royal expense. Each of the seven Indian nobles was to be costumed in a blue velvet sleeveless coat, a doublet of yellow damask and a cap and breeches of scarlet cloth. Each was to have two fine shirts, leather gaiters and shoes decorated with ribbons. The less important Indians were to have sleeveless coats of yellow cloth, doublets of white cotton fustian, scarlet caps and capes of mulberry-colored cloth.

In his concern for the Indians and in a dozen other things — provisions for the founding of churches, monasteries, schools — Charles was amenable to the suggestions of Cortés. He flattered Cortés with attention; when the latter fell ill in Toledo, the king visited him at his bedside, a most unusual condescension. He offered him knighthood in the noble military order of Santiago (an offer that Cortés apparently declined, thinking that anything other than the rank of knight-commander was inappropriate). He reaffirmed his earlier appointment of Cortés as captain-general of New Spain and extended his command to include as yet unfound or unexplored coasts and islands of the South Sea, and the right to one-twelfth of whatever he should conquer in that area. He appointed him Marquis of the Valley of Oaxaca, a broad and fertile region that included twenty-two towns spread over much of south-central Mexico. The grant included twenty-three thousand vassals — a figure Cortés later insisted was twenty-three thousand households, which would have increased the figure about fourfold. Cortés was outwardly appreciative of these grants and honors and was most attentive to the king. When Charles departed for Bologna, where he was to be crowned Roman emperor more than a decade after his election to that post, relations between the two, conquistador and king, appeared to be most cordial.

Still Charles withheld from Cortés the one thing closest to his heart: a restoration of his position as governor of New Spain or, better yet, appointment as viceroy for that vast region. But whatever disappointment Cortés felt was hidden in a flurry of other activity. He sent agents and a company of Indian acrobats and jugglers to perform for Pope Clement VII in Rome. The pope was so pleased that the agents had no trouble obtaining a bull legitimatizing three of Cortés's bastard children, including his favorite, Martín, the son of Marina.

He arranged for his recently widowed mother to travel with him to New Spain. And finally, there was his marriage to Doña Juana. Details of the ceremony are lacking, and it is possible that the event may have been performed by proxy several years earlier, while Cortés was still in New Spain, with only the celebration and consummation delayed. Given the circumstances, it might have been a marriage of little more than convenience. But, Cortés must have been delighted to discover, Doña Juana de Zúñiga was not only well-born. She was also young, beautiful, charming and much admired in royal and aristocratic circles. She came to marriage with a substantial dowry — which Cortés was to find most useful in future ventures — and she was soon to demonstrate a whole-hearted devotion to her husband's interests and welfare.

Cortés embarked on the marriage with more seriousness than he usually devoted to matters of the heart. He presented the fabulous green gems to Doña Juana as a wedding gift. This was remarkable because, while he had been at court, Elizabeth of Portugal, Charles's queen, had indicated that she would like them for herself. Eager as Cortés usually was to cultivate royal favor, he must have been tempted to dispose of the jewels where they would do him the most good. But he kept to the original plan of presenting them to his bride.

If Queen Elizabeth was annoyed at this rebuff, she carefully concealed it. In royal documents that she dictated in her husband's absence she referred to Cortés glowingly as "the governor, Don Hernán Cortés, Marques del Valle," honoring him with the title the king had refused to confer. In orders dated at Toledo April 5, 1529, directed to "councils, justices, aldermen, gentlemen, seal-keepers, officials and good men of all the cities, towns and places of these our kingdoms and principalities," she stated that Cortés was leaving Toledo to be married in the Villa de Bejar, after which he and his bride would embark from Seville for New Spain. Cortés and his wife were to be provided with good lodgings, not merely those of an inn or tavern, and whatever else they might need; for Cortés was valued as a royal servant and it was the queen's wish that he be favored and respected. Similar orders were directed to royal authorities in Española and Cuba, in which Cortés was expecting to stop en route.

As usual when Cortés was absent, there was virtual anarchy in New Spain. A royal *audiencia,* a judicial-administrative body, had been appointed in 1528, the same year that Cortés returned to Spain, to look into the incredible tangle of charges and countercharges concerning Cortés and his conduct. This investigating body, composed of four lawyers, arrived in December of that year. It was to be presided over by Nuño de Guzmán, who had come out with Ponce de León and was serving as acting governor of Pánuco. They were, at first, accommodated in Cortés's palatial home, and there two of them soon died of pneumonia (Bernal Díaz cynically remarked that Cortés was fortunate to be out of the country; otherwise he would have been blamed for their deaths). The survivors, Juan Ortiz de Matienzo and Diego Delgadillo, with Nuño de Guzmán, proceeded on a course designed to (1) enrich themselves, (2) thoroughly discredit Cortés, (3) persecute those who supported Cortés and (4) should Cortés return from Spain, to drive him into open rebellion for which he could be punished.

It was providential for Cortés, for Mexico and Spain that there was one man on hand to observe the excesses of the *audiencia* and report them to the Council of the Indies. This was Fray Juan de Zumárraga, Bishop of Mexico, who had come out with the *audiencia* with a special charge to protect the Indians. Zumárraga reported that Guzmán, in his position as acting governor of Pánuco, was enslaving Indians, branding them and sending them by shiploads for sale in the Indies. Also that the members of the *audiencia* were seizing the prettiest Indian women as concubines. Guzmán had insisted on eighty for himself and had rebuffed the bishop's inquiry with the declaration that this was a private affair. Furthermore the *audiencia* had taken particular vengeance on Pedro de Alvarado, recently returned from Spain, for having spoken there in defense of his old captain, Cortés. Alvarado was deprived of all his goods and thrown into prison with irons on his feet. Zumárraga's report brought quick action in Spain. A new *audiencia* was appointed, and Cortés was urged to return to Mexico as soon as possible.

Cortés would have preferred to wait for the second *audiencia* to arrive, but because of the size of his entourage and the expense of supporting it, he could not delay. He landed in Veracruz on July 15, 1530 (and the second *audiencia* did not sail from Seville until forty days later). Cortés and his party were greeted by throngs of unhappy Spaniards and a multitude of Indians, all eager for deliverance from the unjust administration of the *audiencia*. Cortés, now a grandee of Spain, had himself publicly declared captain-general of New Spain and promised to make matters right. But the *audiencia* ordered Indians not to serve him and Cortés to stay away from the capital. Cortés was to claim later that the withholding of the help and services of the Indians caused the deaths of two hundred persons in his entourage, including his mother. He settled disconsolately in Texcoco, across the lake from the capital, and waited for the arrival of the second *audiencia*.

The second *audiencia* arrived early in 1531. Its members were sober, astute and dedicated men. Cortés was invited to return to the capital, and steps were taken for the formal purchase of his palace, which the *audiencia* occupied. The two earlier auditors, Matienzo and Delgadillo, were charged and imprisoned. Insofar as possible, the properties which had been stolen from Cortés were restored to him, and he was treated with both respect and honor. Nuño de Guzmán, who had proved himself to be perhaps the cruelest of all the Spaniards in New Spain, hastened to launch a bloodthirsty expedition into Jalisco, which he proclaimed to be a new province, New Galicia. He thus escaped the authority of the *audiencia* of New Spain.

A lesser man than Cortés would probably have been content. The marquisate that he had been granted made him a tremendously rich and powerful man. It stretched from ocean to ocean and was perhaps one-fourth as large as Spain itself. Although he continued to argue with authorities about the number of vassals allotted to him, the twenty-three thousand that he had were far more than enough to support him and his descendants in regal style. He had gold and silver mines, plantations of cotton and sugar cane, mills, orchards, grazing lands. In Cuernavaca he built a cathedral and, for himself, a turreted palace, which he apparently hoped to occupy as viceroy of New Spain when the king was ready to make such an appointment — an appointment Cortés had urged the king to make.

But while waiting for an upward turn in his fortunes Cortés was unable — or at least unwilling — to be satisfied with the role of feudal lord. More and more his attention was attracted to the South Sea — that is, the Pacific — as a promising field for his talents.

As early as 1526, in response to an order from the king, he had sent the little fleet of ships he had built at Zacatula across the ocean to find a route to the Moluccas, or Spice Islands, and to search for Spaniards who had been lost on exploratory voyages. He placed a kinsman, Alvaro de Saavedra, in command. Saavedra died during the expedition and the survivors fell into the hands of the Portuguese, with whom Spain was debating ownership of that part of the world. It was an inconclusive sort of venture — and one which must have put a strain on Cortés's finances.

The royal commission given him in 1529 to explore the South Sea, plus Pizarro's successes in Peru, led Cortés into more shipbuilding and exploration ventures. In 1532 he fitted out two vessels at Acapulco. One of them sailed west and was never seen again; the other landed on the coast of Jalisco — in the territory of the unfriendly Nuño de Guzmán — after an unproductive and inconclusive voyage. Cortés built two more vessels, which set out from Tehuantepec in 1534. One of them found a useless desert island, probably in the Gulf of California, and returned to Tehuantepec, having accomplished nothing. The pilot of the other ship, Fortún Jiménez, murdered the captain, took charge himself, and sailed into a harbor in Lower California. The Spaniards found a quantity of pearls, but they also brutalized the very primitive natives, inciting them to an uprising, in which most of the Spaniards were killed. But there were enough survivors to bring back to mainland Mexico tales of rich pearl fishing grounds.

Cortés, who for more than a decade had heard tales of an island somewhere to the west of Mexico inhabited by women warriors who wielded weapons of solid gold, decided to head the next expedition himself. With four vessels and three hundred men — some of whom brought their wives, intending to settle — he sailed from the coast of Sinaloa early in 1535. On May 1 he reached the bay where the Spaniards had been massacred. He named the place Santa Cruz (now La Paz, Baja California). He formally claimed the country, dictating a declaration to the notary who accompanied him: "He took and seized in the name of Your Majesty the tenancy and possession of said newly discovered land . . . and of all else that is afterwards made known or falls within those territories and demarcations . . . as a beginning to pursue the discoveries, conquests and populating of them . . . he commanded the people to have him as governor for Your Majesty. . . ."

The proclamation was a sad and windy epitaph to a distinguished career of conquest. The natives of this new land were few, poor and abysmally backward. Aside from the few pearls to be had in the nearby gulf, the resources of the country were nil. The infant colony could be supported only by importing food and goods from mainland Mexico. The distance was not great, but the gulf waters were treacherous, with vicious currents, dangerous shoals and frequent storms. Cortés sent out ships on exploring missions along the gulf. But his prolonged absence from Mexico brought urgent messages from both his wife and the newly appointed viceroy of New Spain, Don Antonio de Mendoza. After a year Cortés gave up and returned to Mexico. His ships and voyages in the Pacific had cost him some two hundred thousand ducats (approximately one million dollars). Doña Juana's dowry of ten thousand ducats had gone into the ventures, and so had much money that Cortés could ill afford to spend. He filed a claim on the royal treasury for his expenses, and some two years later the claim was allowed — but it was never paid.

His relations with the new viceroy were at first excellent. When a great public celebration was held in Mexico observing the truce of Aigues-Mortes between Charles V of Spain and Francis I of France, Cortés and the new viceroy were genial collaborators in the pageantry. It was noted that the viceroy, a careful man, had a special guard mounted to safeguard his gold and silver plate when he tendered a banquet to Cortés. Cortés, characteristically, took no such precautions when he gave a banquet for the viceroy, and he was said to have lost at least a hundred marks' worth of silver plate to his light-fingered guests. Not only that: he suffered a severe leg wound from a spear thrust during the jousting that was a part of the celebration; he devoted just as much enthusiasm and spirit to mock warfare as he did to the real kind.

Cortés continued to plan expeditions. When word came to Mexico of the supposed existence of the Seven Cities of Cíbola, Cortés was ready to go. But the viceroy denied him the authorization. He would attend to these matters himself.

His only recourse for this and other supposed injustices was to appeal, once more, to the king in person. Early in 1540, with a purseful of gold and the five great emeralds he had presented to Doña Juana, he left again for Spain, accompanied by one legitimate son, Don Martín, and one bastard, Don Luis, whom he had had by a Spanish woman, Antonia Hermosillo.

It was his ultimate attempt to secure royal favor and official blessing. And it was a disappointment almost from the outset. The queen, Elizabeth, had died the previous year, and Charles was distraught over the loss. To make matters worse, his continued extravagances and mounting debts — on which crippling interest rates, as high at 43 percent, had to be paid — were stirring much internal dissension in Spain. Protestantism was spreading rapidly in Germany and Flanders despite all that Charles did to try to eradicate it, and so was political unrest, even in his native city of Ghent. Charles was away from Spain most of the time, and even had he been there it is unlikely that he would have had time or patience for the troubles of Hernán Cortés. The advisers whom he had left in charge in Spain were sympathetic and respectful with Cortés, but it was clear they could take no action in the king's absence.

In the fall of the following year Charles launched a massive attack against Hassan Aga, a henchman of the Barbary pirate Barbarossa, in Algiers, with four hundred fifty ships, sixty-five galleys, twelve thousand sailors and twenty-four thousand troops, recruited from Germany, Italy and Spain. Hassan Aga's garrison consisted of no more than seventy-five hundred men, and the assault by the Christians seemed assured of success. But at the critical moment the attacking forces were struck by a storm, which rendered the guns useless and wrecked a large part of the supporting fleet. This was on October 25, 1541 — and it was the most disastrous single event of Charles's reign. At least one hundred fifty ships and possibly as many as twelve thousand men were lost. After a council of war it was decided to withdraw.

The Spanish contingent had included Spain's most illustrious soldiers, including Hernán Cortés, who was accompanied by his two sons. Their ship was wrecked on the Algerian coast, and while they escaped, Cortés lost everything he had with him, including the famous emeralds.

But this was not the greatest blow to Cortés. Although he was Spain's most noted soldier, he was not invited to participate in the council of war. When the decision to withdraw was announced, he sent word to the king that if he could be entrusted with a small body of the troops he would remain behind and overcome Hassan Aga and the Moors; it would be an easier task than had been the overthrow of the Aztec empire. His offer was laughed at and ignored.

This was the end of Hernán Cortés. Although he lived six more years, his pride and self-esteem had been dealt a mortal blow. Voltaire, in his *Essai sur les Moeurs,* told a story of a subsequent meeting between Cortés and his king. The old captain, desperately seeking a word with the monarch, clutched the steps of the royal carriage, like a common beggar, asking to be heard. Charles must have known who the gray-beard was, but he rudely asked, "Who are you?"

"Sire, I am a man," said Cortés, "who has given your Majesty more provinces than you possessed cities."

The king ordered his coachman to drive on. The anecdote is probably fictional, but the spirit is true.

One letter that Cortés addressed to his king in these last years survives — although there must have been many more. Its bitterness is tempered with sadness. Writing from Valla-dolid on February 3, 1544, Cortés told his king, "I thought that having labored in my earlier years I would enjoy rest in my old age; so it was that I spent 40 years with little rest, eating bad food and, in good times and bad, bearing my arms in uphill struggles, putting my person in danger, spending my substance in service to God, bringing sheep to His corral in remote and unknown provinces, glorifying and expanding the name and patrimony of my King, winning and subjecting to his rule and scepter many great kingdoms and principalities of barbarous nations and peoples, won by my own efforts and expense without any assistance whatsoever." To make matters worse, he said, he had been "obstructed by many covetous rivals who, like leeches, have satiated themselves on [his] blood." But after a final plea to the King to see that justice was done, he concluded: "I must return to my house; I am no longer of an age to loiter in inns. I must retire and settle my account with God, for it is a long one, and I have little life left to put my case before Him. . . . It is better to lose one's wealth than one's soul."

Finally, in 1547, he gave up and prepared to return to Mexico. But his life was ebbing fast. He withdrew to the small town of Castilleja de la Cuesta, outside Seville, and took lodging in the simple home of a magistrate. And there, on December 2, 1547, the greatest of the conquistadors died.

SEVENTEEN

Epilogue

Fifty-two days before his death, Cortés was in Seville, preparing to return to New Spain. He suspected, correctly, that he would not live to make the voyage, and he drew up his will, settling his "account with God," as he put it in his last letter to the king.

There were the usual expressions of pious mortality: "I . . . Marquis of the Valley of Oaxaca, Captain General of New Spain for the Caesarian Majesty of Emperor Charles . . . being ill but in such free and sound judgment with which it has pleased God to endow me, fearing death, as is natural in every creature, and desiring to prepare myself . . . do for the good of my soul and the peace and discharge of my conscience, execute and recognize this document. . . ." He gave detailed instructions for his burial, first in Spain and later in New Spain, and provided for the establishment of a hospital in the city of Mexico and a monastery and seminary in Coyoacan, in addition to the monastery and churches he had already established.

But with these conventions out of the way, Cortés proceeded in a manner completely alien to the righteous self-justification that had colored — and probably distorted — most of his earlier letters and the romantic bombast with which he had addressed his followers in the great adventure.

He admitted that he had done things which, in justice, should be corrected after his death. He acknowledged that much of the property he held could rightfully be claimed by others. In such cases he ordered restitution to the rightful owners and compensation to them for the use he had had of the properties. He specifically included Indian lands that he had used for orchards, vineyards and cotton fields.

He was troubled by the problem of slavery. A postmortem inventory of his holdings would later show that he owned only twenty-five slaves — sixteen Indians and nine Negroes — but it is likely that he had held many more. In his will he observed, "There have been many doubts and opinions as to whether it is permitted with good conscience to hold . . . slaves, whether captives of war or by purchase. . . . I direct my son and successor . . . and those who may follow him, to use all diligence to settle this point for the peace of my conscience and their own."

There was also a difference of opinion, he noted, on the Spaniards' right to exact tribute and services from Indian vassals. "This matter shall be investigated," he ordered, "and if it appears that I have received more . . . than belonged to me, those natives shall be paid and indemnified in all that it shall appear they may justly claim."

He made generous provisions for his family. His widow, Doña Juana, in addition to other bequests, was to be reimbursed for the ten-thousand-ducat dowry Cortés had spent on his inconsequential Pacific explorations. He had sired six children by Doña Juana, four of whom were still living. His one legitimate son, Martín, born in Cuernavaca in 1532, was heir to the marquisate and the bulk of the estate. Each of the three legitimate daughers, María, Catalina and Juana, born between 1533 and 1536, was to be provided with a large dowry.

More remarkable in an age when illegitimate children were

commonly forgotten or ignored was Cortés's concern for his *hijos naturales*. Five of them were accounted for in the will. He made particularly handsome bequests to his illegitimate daughter Catalina, child of a Cuban Indian woman, Leonor Pizarro (for whom Cortés had arranged baptism and christening with his mother's family name). During his lifetime he had given Catalina various properties. In his will, in addition to providing her with a dowry and other bequests, he admitted that he had continued to receive the income from these properties, and he ordered that she be reimbursed. In later years the widow, Doña Juana, managed to force Catalina to sign over these properties to the legitimate family and then had her taken to Spain, where she was placed in a convent. The marriage to the son of Francisco de Garay, which her father had arranged for her, had never taken place.

He also made substantial bequests to his other illegitimate daughters. One was Leonor, whose mother was the eldest daughter of Moctezuma. Leonor's daughter, Isabel de Tolosa Cortés Moctezuma, married Juan de Oñate, colonizer and first governor of New Mexico. There was also provision for a third illegitimate daughter, whose mother was identified only as an Aztec princess.

Cortés also provided large pensions for his two bastard sons: Martín, whose mother was Marina, Cortés's interpreter and mistress, and Luis, his son by Antonia Hermosillo. In the case of Luis, however, Cortés, on the day of his death, added a codicil to his will canceling the thousand-ducat annual pension he had originally bequeathed him. The reason was believed to be that Luis planned to marry the niece of an old enemy of Cortés, Bernaldino Vázquez de Tapia.

The three sons — the legitimate Martín and his half-brothers Martín and Luis — spent most of their early and middle years in Spain. The second marquis was close to Philip II of Spain, who succeeded to the throne when his father, Charles V, abdicated in 1556. He married the king's niece, Doña Ana Ramírez de Arellano. When the marquis returned to Mexico the king granted him unrestricted use of the marquisate of the Valley, which meant, by this time, an income of eighty-six thousand pesos yearly.

When the marquis arrived in Mexico, accompanied by his half-brothers Martín and Luis, he was fervently greeted by the *criollos* (creoles: that is, Spaniards born in the New World). There were elaborate ceremonies, processions, feasts, tournaments and a pageant depicting the arrival of the first marquis in Tenochtitlan and his meeting with Moctezuma.

It was more than a matter of sentiment and nostalgia. The creoles were fearful that the Spanish crown was going to deprive them of the *encomiendas* their grandfathers had won in New Spain. The second viceroy, Don Luis de Mendoza, a just man, showed strong sympathies for the Indians, an attitude the creoles regarded as treacherous. The second marquis was a living symbol of what they regarded as their rightful inheritance. Soon there were plans afoot for a rebellion against the viceroy and the crown, with the aim of establishing the second marquis as king of an independent Mexico.

The movement became known as the Conspiracy of 1565. If the second marquis had been the man his father was, the movement might have succeeded. But Don Martín was, in contrast, a weak and arrogant man, susceptible to flattery, vacillating, and cautious almost to the point of cowardice. He survived, as did the half-brothers, who were both involved in the plot — although the other Don Martín, Marina's son, was subjected to severe torture — which was better than the beheading that many of the plotters suffered. All three of Cortés's sons were exiled from the country their father had conquered and the nation he had created.

By the time of this conspiracy, the woman who had made the conquest possible was long dead. During her last service to Cortés — as interpreter during the march to Honduras — Marina had a reunion with her mother, who had sold her to slave traders years earlier. The mother, by now christened Marta, and her son Lázaro were brought to the Spaniards' camp at Coatzacoalcos. The older woman and the young man were trembling with fear. But Marina forgave her mother, treated both of them with kindness and loaded them with gifts. Beyond the fact that she bore a daughter to Juan de Jaramillo, little more is known of Marina, except that her death occurred around 1540. In 1605 Don Fernando Cortés, son of the illegitimate Don Martín Cortés, addressed a memorial to the Spanish court detailing the services his grandmother had performed during the conquest. With that the record, official and otherwise, of Malinali, Marina or Malinche ends. The name Malinche, however, unhappily lives on in Mexican slang as a pejorative term for persons who betray their nation and heritage.

A few of Cortés's companions survived the violent struggles for domination of the New World to achieve prominence or wealth or both. But lasting fame came to few — and one of the few was Bernal Díaz del Castillo, a common foot soldier during most of the conquest. Bernal Díaz, seven years younger than Cortés, began, at the age of eighty-four, to dictate his "true history." Blind and deaf, he was then a magistrate in Guatemala. Only five of the men who had originally set out with Cortés were still alive, and all were ailing and poor. Bernal Díaz was angered by the lustrous, eloquent accounts of the conquest that had been written, principally by the one published by Francisco López de Gómara, Cortés's chaplain and secretary in his last years, a man who had never set foot in Mexico, never endured the perils and hardships of the conquest. Apologizing for his lack of polish, Bernal Díaz began: "That which I have myself seen and the fighting I have gone through, with the help of God I will describe quite simply as a fair eyewitness without twisting events one way or another. . . . I have gained nothing of value to leave my children and descendants but this my true story. . . ."

Bernal Díaz's account was homely, graphic, inaccurate in many things, but it had the ring of truth and won for its author a place as one of the greatest amateur historians of all time. In his book the Spanish destinies were not always guided by divine will but often by sheer chance. The conquistadors were not always gallant and brave. They were just as often avaricious, unscrupulous and frightened to death.

Cortés, in Bernal Díaz's eyes, was not the paragon that Cortés himself described in his letters to the king nor the imperious figure that emerged in Gómara's sophisticated rhetoric. Instead he was an affable leader who often sought the advice of his followers but one who did not hesitate to cheat everyone when it came time to divide up the spoils. He was not a man who always found comfort in his own righteousness; instead he was a man assailed by doubts and troubled by his conscience. And, while he was guilty of much ruthlessness and needless cruelty, he was, by far, more thoughtful and more considerate of the Indian masses of Mexico than were most of his companions. The proof was found in the warmth with which he was welcomed back to Mexico by the Indians after each of his absences, absences during which chaos ruled.

Such a balanced view of Cortés was not to be common in Mexico. To the white aristocracy and much of the Catholic clergy he was the heroic founder, the bringer of Christianity, salvation and a Spanish heritage which they considered glorious. To the masses, he was the principal villain in the long years of Spanish colonial repression.

By the time Mexico threw off Spanish rule, three centuries after the conquest, there was widespread hatred of Cortés and a thirst for vengeance. Cortés's remains became the target of public fury. The remains had, by this time, traveled widely. In 1566 they were disinterred in Seville and transported to Mexico, where they were reburied in the church of San Francisco in Texcoco, where his mother and an infant son had been buried long before. When his grandson, Don Pedro Cortés, the fourth marquis, died in 1629, the remains were again disinterred so that they could be buried in the convent of San Francisco in Mexico City in the same mausoleum with the body of the grandson. In 1791 the viceroy approved a proposal to dig them up once more and deposit them in the church of the Hospital of Jesus — another of the institutions Cortés had founded. A mausoleum surmounted with a bust of Cortés was built and the remains were deposited there in 1794. Cortés partisans and descendants must have hoped that this would be the last move.

But in 1823, after the war of independence, the bodies of many insurgent heroes were brought to Mexico City for burial. The occasion produced a public demand that the mausoleum — a monument of sorts to Spanish domination of Mexico — be destroyed and that Cortés's bones be publicly burned on September 16, the anniversary of the launching of the struggle for independence. On the eve of the anniversary ecclesiastical authorities ordered the mausoleum dismantled and the Cortés remains reburied in a secret place. This was done in the presence of a representative of the Duke of Terranova, fourteenth Marquis of the Valley.

The secret was well kept, and it was for many years supposed that whatever was left of the mortal remains of Cortés had been taken for safekeeping to Palermo, Italy, where the title was now held. The bust and the gilded bronze arms that decorated the tomb were taken there. In 1929, however, Prince Antonio Pignatelli, the latest Marquis of the Valley, said that the remains were believed to still be in the Hospital of Jesus. But no one knew exactly where.

Finally, in 1946 a party of antiquarians, after careful examination of some old records — the source of which they refused to reveal — dug a hole in the wall near the altar of the old abandoned church of the Hospital of Jesus. There, in a blue-painted niche behind a stone slab, they found a small casket covered with gold-trimmed black velvet and decorated with a gold cross. The outer casket was made of lead. Inside it was a wooden casket. Within it and protected by another sheath of lead was a glass urn decorated with gilded metal. In the urn were a skull wrapped in a cambric handkerchief, and a clutch of bones in a white sheet bordered with black lace.

There was also a metal tube containing a notarized statement that these were the remains of Hernán Cortés, who had died in Castilleja de la Cuesta, Spain, just one year short of four hundred years before.

The president of Mexico ordered that the bones be reburied in the same place, and that the old church — used only by pigeons for the twenty years since the bitter anticlerical aftermath of Mexico's twentieth century revolution — be made a national monument. It was an appropriate and dignified decision, but it did not calm the political storm that the discovery had brought about. Religious and conservative elements demanded public observance of Mexico's debt to Spain and the man who had eradicated Mexico's ancient heathen cults. Liberal and radical elements denounced the conquistador's brutality.

Nor was the controversy confined to public utterances. A doctor and a historian, both hispanophobes, examined the bones at the Hospital of Jesus before they were reinterred. Out of their observations came a revised version of Cortés's physical appearance that was considerably at variance with the descriptions of eyewitnesses. He was, they said, a virtual monster, a fraction more than five feet four inches tall, hunchbacked, with a pointed head, a narrow bulging forehead, beaked nose, receding chin, pigeon breast, bowed legs.

The painter Diego Rivera, an admirable craftsman of somewhat eccentric political passions, used this "rectification" in two portraits of Cortés in his fresco "Colonial Domination" on the balcony of the National Palace in Mexico City, a work completed in 1951. He took the liberty of adding a few "corrections" of his own invention: ashy skin, crossed eyes and what appears to be a syphilitic lesion high on the bald forehead.

Not long after the discovery in the Hospital of Jesus it was claimed that the remains of Cuauhtemoc, the last and most heroic of the Aztec rulers and Cortés's victim on the melancholy march to Honduras, had been found. The location was the remote village of Ixcateopan, southwest of Taxco in the state of Guerrero, roughly six hundred air miles from the scene of his death — and much farther for anyone who had to traverse it on foot. Of this latter discovery, standard Mexican reference works say mildly that the matter has been much argued about. Was it an authentic, historically sound discovery? Or only a manifestation of vigorous nationalism and enduring hatred of the man who created the nation? No one knows.

Cortés's fame swung, pendulumlike, between grandeur and humiliation in his lifetime. More than four centuries after his death the pendulum continues to swing.

Suggestions for Further Reading

In a short book intended only to introduce a fascinating character and an exciting period of history it is useless to burden the reader with a list of all sources, documents and archives consulted, arrayed as were the jawbones of enemies on the arms of victorious Mayan warriors.

However, certain debts should be acknowledged. And, in the hope that this essay in biography may inspire some readers to dig further into the man, the time and the tumultuous circumstances, a few of the more interesting, helpful and easily available sources should be indicated.

Any consideration of the life, career and character of Hernán Cortés (also known as Hernando or Fernando Cortéz) should begin with three basic sources.

The first is the series of five letters that Cortés wrote (or, in the case of the first one, ordered written) to his king, Charles V, regarding his adventures in Mexico. The letters have the common fault of autobiography. Cortés is generous with himself and his deeds and is often negligent of his companions. There is much that borders on self-glorification, much partisan argument, much pleading of his own case. His facts are occasionally awry, and his phonetic rendition of Indian names is both reckless and inconsistent. Still, it is an exciting, terse story of one of the greatest and most hazardous military undertakings of all time. Best of the available editions, and the one used frequently in this book, is *Hernán Cortés: Letters from Mexico,* translated and edited by A. R. Pagden (New York: Grossman Publishers, 1971). The text is highly readable and the annotation superb.

Equally interesting and helpful — and extensively used here — is the biography of Cortés written by the man who served him as secretary and chaplain in his late years, Francisco López de Gómara, published as *Cortés, The Life of the Conqueror by his Secretary,* translated and edited by Lesley Byrd Simpson (Berkeley and Los Angeles: University of California Press, 1964). López de Gómara's text is of considerable literary merit, with much color and drama, and Simpson's translation is careful, literate and skillful. Its principal drawbacks are that López de Gómara was never in Mexico, was not an eyewitness of the events he describes so graphically. Instead, he acquired all his information from others — principally Cortés himself, who was hardly an unbiased source.

López de Gómara made another contribution, unwittingly. His writings on Cortés and the Conquest so infuriated the veteran conquistador, Bernal Díaz del Castillo, that, although he was by that time an old man, almost blind and deaf, he proceeded to write (or dictate) his own version of Cortés and the Conquest from the viewpoint of a man who had been there and suffered it all. He attempted to tell what truly happened, not hearsay glossed over with polished rhetoric. He had little to show for his part in the Conquest, and the book, which he did not live to see published, was to be his only legacy to his heirs. His version was inaccurate in many things; memory plays tricks on the aged. He, no less than Cortés, tended to claim more credit than was perhaps due him. As with Cortés, his version of Mexican words and names was mystifying. But, although he sometimes displayed petty jealousy, he was a sharp-eyed observer, and there was a sturdy, earthy candor in all that he wrote. He was sensible to both martial skills and human foibles, including his own. He did not hesitate to give his Indian foes high marks as valiant and able warriors. He was one of the have-nots of the Conquest, but he did not let bitterness diminish his pride in what he and his companions had accomplished. His "true story" is essential to any comprehension of the subject. Probably the best current edition is *The Bernal Díaz Chronicles,* translated and edited by Albert Idell (Garden City: Doubleday, 1957), and it has been much relied on here. But the Idell version takes the story only through the fall of Tenochtitlan. For the rest of the story — and the rest is both interesting and important — the earliest English translation, by Maurice Keating (or Keatinge), published in London in 1800, is now available in a facsimile edition published by University Microfilms, Ann Arbor, Michigan, 1966.

For the *relaciones* or memoirs of other participants, *The Conquistadors,* edited and translated by Patricia de Fuentes (New York: Orion Press, 1963), provides colorful and helpful commentaries (the accounts of Francisco de Aguilar and Andrés de Tapia, for example) not easily available elsewhere.

The single most comprehensive account of both Cortés and the Conquest is still to be found in William Hickling Prescott's *History of the Conquest of Mexico.* It is a monumental book of compelling style and scholarship, and it is all the more remarkable because during the period of researching and writing Prescott was blind or nearly so. Ever since its publication in 1843 it has been a landmark in American literary-historical achievement. For the person interested in history in general and Cortés in particular it is most highly recommended.

The serious student who does not have access to primary source materials is urged to consider *The Rise of Fernando Cortés,* by Henry R. Wagner (Berkeley: The Cortés Society, Bancroft Library, 1944). Although Mr. Wagner's prose is graceless and perplexing, his research and annotation are prodigious, his critical analyses are penetrating, and he had access to many materials that Prescott never saw. He apparently intended to write a second book on the decline and fall of Cortés. It is unfortunate that he was unable to do so.

For a different view of Cortés and his companions and a bitter appraisal of their achievement, the reader is urged to consult the collection of Aztec accounts of the Conquest, translated from the original Nahuatl by Angel María Garibay K. and edited by Miguel León Portilla. There is an excellent English translation by Lysander Kemp, published as *The Broken Spears* (Boston: Beacon, 1962).

Although it quickly becomes a separate and intriguing study, a quick survey of the American peoples and cultures Cortés overcame adds a useful and illuminating dimension. At a bare minimum the serious student should consult:

Ignacio Bernal, *Mexico Before Cortéz,* translation and foreword by Willis Barnstone (Garden City: Doubleday, 1963).

George C. Vaillant, *Aztecs of Mexico* (Garden City: Doubleday, 1953).

Eric R. Wolf, *Sons of the Shaking Earth* (Chicago: University of Chicago Press, 1959).

Nigel Davies, *The Aztecs* (London: Macmillan, 1973).

Index

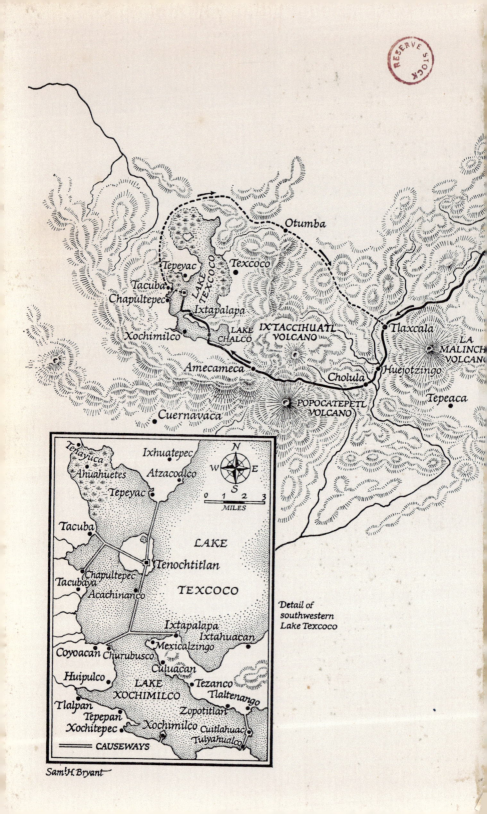

Otumba

Tepeyac   Texcoco

Tacuba
Chapultepec   Ixtapalapa   LAKE TEXCOCO   IXTACCIHUATL VOLCANO   Tlaxcala

Xochimilco   LAKE CHALCO   LA MALINCHE VOLCANO

Amecameca   Huejotzingo

Cholula

Cuernavaca   POPOCATEPETL VOLCANO   Tepeaca

Detail of southwestern Lake Texcoco

Tenayuca   Ixhuatepec
Ahuahuetes   Atzacoalco
Tepeyac

N
W · E
S
0 1 2 3
MILES

Tacuba

LAKE

Chapultepec
Tacubaya   TEXCOCO
Acachinanco   Tenochtitlan

Ixtapalapa
Ixtahuacan
Coyoacan Churubusco   Mexicalzingo
Culuacan
Huipulco
Tezanco
LAKE   Tlaltenango
XOCHIMILCO
Tlalpan
Tepepan   Zopotitlan
Xochitepec   Xochimilco   Cuitlahuac
Tulyahualco

CAUSEWAYS

Sam! H. Bryant